Clustering in Bioinformatics and Drug Discovery

CHAPMAN & HALL/CRC
Mathematical and Computational Biology Series

Aims and scope:

This series aims to capture new developments and summarize what is known over the entire spectrum of mathematical and computational biology and medicine. It seeks to encourage the integration of mathematical, statistical, and computational methods into biology by publishing a broad range of textbooks, reference works, and handbooks. The titles included in the series are meant to appeal to students, researchers, and professionals in the mathematical, statistical and computational sciences, fundamental biology and bioengineering, as well as interdisciplinary researchers involved in the field. The inclusion of concrete examples and applications, and programming techniques and examples, is highly encouraged.

Series Editors

N. F. Britton
Department of Mathematical Sciences
University of Bath

Xihong Lin
Department of Biostatistics
Harvard University

Hershel M. Safer

Maria Victoria Schneider
European Bioinformatics Institute

Mona Singh
Department of Computer Science
Princeton University

Anna Tramontano
Department of Biochemical Sciences
University of Rome La Sapienza

Proposals for the series should be submitted to one of the series editors above or directly to:
CRC Press, Taylor & Francis Group
4th, Floor, Albert House
1-4 Singer Street
London EC2A 4BQ
UK

Published Titles

Algorithms in Bioinformatics: A Practical Introduction
Wing-Kin Sung

Bioinformatics: A Practical Approach
Shui Qing Ye

Biological Sequence Analysis Using the SeqAn C++ Library
Andreas Gogol-Döring and Knut Reinert

Cancer Modelling and Simulation
Luigi Preziosi

Cancer Systems Biology
Edwin Wang

Cell Mechanics: From Single Scale-Based Models to Multiscale Modeling
Arnaud Chauvière, Luigi Preziosi, and Claude Verdier

Clustering in Bioinformatics and Drug Discovery
John D. MacCuish and Norah E. MacCuish

Combinatorial Pattern Matching Algorithms in Computational Biology Using Perl and R
Gabriel Valiente

Computational Biology: A Statistical Mechanics Perspective
Ralf Blossey

Computational Hydrodynamics of Capsules and Biological Cells
C. Pozrikidis

Computational Neuroscience: A Comprehensive Approach
Jianfeng Feng

Data Analysis Tools for DNA Microarrays
Sorin Draghici

Differential Equations and Mathematical Biology, Second Edition
D.S. Jones, M.J. Plank, and B.D. Sleeman

Engineering Genetic Circuits
Chris J. Myers

Exactly Solvable Models of Biological Invasion
Sergei V. Petrovskii and Bai-Lian Li

Gene Expression Studies Using Affymetrix Microarrays
Hinrich Göhlmann and Willem Talloen

Glycome Informatics: Methods and Applications
Kiyoko F. Aoki-Kinoshita

Handbook of Hidden Markov Models in Bioinformatics
Martin Gollery

Introduction to Bioinformatics
Anna Tramontano

Introduction to Computational Proteomics
Golan Yona

An Introduction to Systems Biology: Design Principles of Biological Circuits
Uri Alon

Kinetic Modelling in Systems Biology
Oleg Demin and Igor Goryanin

Knowledge Discovery in Proteomics
Igor Jurisica and Dennis Wigle

Meta-analysis and Combining Information in Genetics and Genomics
Rudy Guerra and Darlene R. Goldstein

Modeling and Simulation of Capsules and Biological Cells
C. Pozrikidis

Niche Modeling: Predictions from Statistical Distributions
David Stockwell

Normal Mode Analysis: Theory and Applications to Biological and Chemical Systems
Qiang Cui and Ivet Bahar

Optimal Control Applied to Biological Models
Suzanne Lenhart and John T. Workman

Pattern Discovery in Bioinformatics: Theory & Algorithms
Laxmi Parida

Python for Bioinformatics
Sebastian Bassi

Spatial Ecology
Stephen Cantrell, Chris Cosner, and Shigui Ruan

Spatiotemporal Patterns in Ecology and Epidemiology: Theory, Models, and Simulation
Horst Malchow, Sergei V. Petrovskii, and Ezio Venturino

Stochastic Modelling for Systems Biology
Darren J. Wilkinson

Structural Bioinformatics: An Algorithmic Approach
Forbes J. Burkowski

The Ten Most Wanted Solutions in Protein Bioinformatics
Anna Tramontano

Chapman & Hall/CRC Mathematical and Computational Biology Series

Clustering in Bioinformatics and Drug Discovery

John D. MacCuish

Norah E. MacCuish

CRC Press
Taylor & Francis Group
Boca Raton London New York

CRC Press is an imprint of the
Taylor & Francis Group, an **informa** business

A CHAPMAN & HALL BOOK

CRC Press
Taylor & Francis Group
6000 Broken Sound Parkway NW, Suite 300
Boca Raton, FL 33487-2742

First issued in paperback 2018

ISBN-13: 978-1-4398-1678-3 (hbk)
ISBN-13: 978-1-138-37423-2 (pbk)

Library of Congress Cataloging-in-Publication Data

MacCuish, John D.
  Clustering in bioinformatics and drug discovery / John D. MacCuish, Norah E. MacCuish.
    p. ; cm. -- (Chapman & Hall/CRC mathematical and computational biology series)
  Includes bibliographical references and index.
  Summary: "This book presents an introduction to cluster analysis and algorithms in the context of drug discovery clustering applications. It provides the key to understanding applications in clustering large combinatorial libraries (in the millions of compounds) for compound acquisition, HTS results, 3D lead hopping, gene expression for toxicity studies, and protein reaction data. Bringing together common and emerging methods, the text covers topics peculiar to drug discovery data, such as asymmetric measures and asymmetric clustering algorithms as well as clustering ambiguity and its relation to fuzzy clustering and overlapping clustering algorithms"--Provided by publisher.
    ISBN 978-1-4398-1678-3 (hardcover : alk. paper)
    1. Bioinformatics--Mathematics. 2. Drug development. 3. Cluster analysis. 4. Computational biology. I. MacCuish, Norah E. II. Title. III. Series: Chapman & Hall/CRC mathematical and computational biology series (Unnumbered)
    [DNLM: 1. Cluster Analysis. 2. Drug Discovery--methods. 3. Computational Biology. QV 744]

  QH324.2.M33 2011
  572.80285--dc22   2010025741

Contents

Supplementary Resources Disclaimer

Additional resources were previously made available for this title on DVD. However, as DVD has become a less accessible format, all resources have been moved to a more convenient online download option.

You can find these resources available here: https://www.routledge.com/9781439816783

Please note: Where this title mentions the associated disc, please use the downloadable resources instead.

List of Figures

List of Tables

Preface

Drug discovery is a complicated process, requiring skills and knowledge from many fields: biology, chemistry, mathematics, statistics, computer science, and physics. Almost without exception, even the simplest drug interactions within the body involve extremely complicated processes. Professionals in numerous sub-fields (cell biologists, medicinal chemists, synthetic chemists, statisticians, etc.) make up teams that, in sequence and in parallel, form a network of research and process that leads — hopefully — to new drugs and to a greater understanding of human health and disease, as well as, more generally, the biological processes that constitute life. The advent of computer technology in the latter half of the 20th century allowed for the collection, management, and manipulation of the enormity of data and information flowing from these endeavors. The fields of cheminformatics and bioinformatics are the outgrowth of the combination of computing resources and scientific data. Modern drug discovery increasingly links these two fields, and draws on both chemistry and biology for a greater understanding of the drug–disease interaction.

Cluster analysis is an exploratory data analysis tool that naturally spans many fields, and bioinformatics and cheminformatics as they relate to drug discovery are no exception. Exploratory data analysis is often used at the outset of a scientific enquiry; experiments are designed, performed, and data are collected. The experiments that collect observations may be entirely exploratory, in that they do not fit under the more rigorous statistical confines of *design of experiments* methodology, whereby hypotheses are being tested directly. Experiments can simply be designed to collect observations that will be *mined* for possible structure and information. Namely, the scientific process is often not quite as straightforward nor as simple as a recanting of the scientific method might suggest; hypotheses do not necessarily appear whole cloth with a few observations. There are a great many fits and starts in even understanding what questions might be asked of a set of observations. Intuitions about observations may or may not be forthcoming, especially in the face of enormous complexity and vast numbers of observations. Finding structure in a set of observations may help to provide intuitions based on known facts that lead to hypotheses that may bear fruit from subsequent testing — where the scientific method earns its keep. Cluster analysis is a set of computational methods that quantify structure or classes within a data set: are there groups in the data? This effort is in turn frought with the possible misunderstanding that any structure found through these methods must be something other than a random artifact. The observations may not mean much of anything!

This book is an exploration of the use of cluster analysis in the allied fields of bioinformatics and cheminformatics as they relate to drug discovery, and it is an attempt to help the practitioners in the field to navigate the use and misuse of these methods—that they can in the end understand the relative merits of clustering methods with the data they have at hand, and that they can evaluate and inspect the results in the hope that they can thereby develop useful hypotheses to test.

How to Use This Book

The first three chapters set the stage for all that follows. Chapters 4, 5, and 6 can be used to provide a student with sufficient detail to have a basic understanding of cluster analysis for bioinformatics and drug discovery. Such a selection of the text could be used as an introduction to cluster analysis over a 2- or 3-week period as a portion of a semester-long, introductory class. Students or practitioners interested in additional and more advanced methods, and subtle details related to specific types of data, would do well to cover the remaining chapters, 7 through 11. This would represent a more advanced, semester-long course or seminar in cluster analysis.

For supplementary materials, a great deal can be obtained online, including descriptions of methods, data, and free and open source software. However, for more extensive treatment of cluster analysis in general there are numerous texts of varying quality published over the past 40 years, the most outstanding of which, though somewhat outdated from a computational point of view, is *Algorithms for Clustering Data* by Jain and Dubes. It is, however, out of print and difficult to find. A more accessible and obtainable text is *Finding Groups in Data* by Kaufman and Rousseeuw. Algorithms in this latter text can be found in R and S-Plus. A passing familiarity with R or S-Plus is extremely handy in exploring cluster algorithms directly. Students and practitioners would do themselves a significant service by learning the rudiments of one of these nearly identical languages. Readers having some exposure to interpreted languages such as Python, or those found in software packages such as Mathematica, Maple, or MATLAB, should have little trouble picking up the syntax and constructs of R or S. Note: all of the figures in the text are in grayscale, but many of the more complex figures have color representations that can be found on the DVD included with this book.

Acknowledgments

Cluster analysis requires well-engineered software. In large measure we have Mitch Chapman to thank for a great deal of the software engineering of the software written by ourselves, including refactoring, testing, maintaining, revision control, build processes, and otherwise, above all, improving and performance enhancing. We would like to thank Bernard Moret for suggesting the outstanding text by Jain and Dubes, *Algorithms for Clustering Data*, as a first start in cluster analysis. John Bradshaw's keen interest in drug discovery and cluster analysis in particular among his many areas of expertise helped to motivate us over the years through discussions and collaborations with him. His work on asymmetry led to our exploration and experimentation of this interesting topic. Discussions regarding discrete measures gave impetus to our study of discrete measures and ultimately to the more general exploration of clustering ambiguity. Thanks to Peter Willett for his life's work in clustering in cheminformatics, which continues to inspire and engage us. Likewise, Yvonne Martin alerted us to various portions of the literature that was instrumental in exploring the development of specialized leader/exclusion region algorithms for large scale clustering. John Blankley helped to identify problems in our leader algorithm development. Elizabeth Johnson helped to keep us on the straight and narrow regarding computational complexity. Michael Hawrylycz helped us with his broad knowledge of bioinformatics applications and mathematics. William Bruno contributed his insights into phylogenetic clustering. Thanks to the BioComputing group at the University of New Mexico for pushing the limits of our clustering software. Also, we give thanks to Jeremy Yang for a continual flow of useful additions. We acknowledge Dave Weininger and Christine Humblet for supporting initial work on cluster analysis for compound diversity, and Anthony Nicholls for supporting expansion of these techniques into conformational space.

We would like to thank our family and friends who have endured our seclusion with patience and support. And most of all we would like to thank our children, Megan and Isaiah, for putting up with having two scientists for parents, and allowing us to write, what they call derisively, *"only a math book."*

John D. MacCuish
Norah E. MacCuish
Santa Fe, New Mexico

About the Authors

John D. MacCuish received his MS in computer science from Indiana University in 1993, where his research interests were graph theory, algorithm animation, and scientific visualization. After graduating from Indiana University, he worked for 3 years as a graduate research assistant at Los Alamos National Laboratory, while pursuing a Ph.D. in computer science at the University of New Mexico. At Los Alamos, he worked on numerous projects, such as fraud detection, a software workbench for combinatorics, and parallel implementations of cluster algorithms for image processing. In 1996, John left the laboratory and the Ph.D. program at UNM to join the software and consulting fields as a scientist working on data mining projects across industries. He developed software and algorithms for data mining and statistical modeling applications, such as anomaly detection and automated reasoning systems for drug discovery. In 1999, he formed Mesa Analytics & Computing, Inc. John has published numerous articles and technical reports on graph theory, algorithm animation, scientific visualization, image processing, cheminfomatics, and data mining. He has also co-authored several software patents.

    Norah E. MacCuish received her Ph.D. from Cornell University in the field of theoretical physical chemistry in 1993. Her educational training included experimental and theoretical work in the areas of ultrafast laser spectroscopy, spectroscopic simulation of fluids, synthetic organic chemistry, and biochemistry. Norah did her post-doctoral work in the pharmaceutical industry in such areas as compound diversity assessment for compound acquisitions and structural diversity analysis of compound leads from high throughput screening. She continued in the pharmaceutical industry as a scientist working on therapeutic project teams involved in combinatorial library design and data integration projects. To further her career in cheminformatics she joined a scientific software vendor, Daylight Chemical Information Systems, in 1997, as a chemical information systems specialist. After a number of years at Daylight, Norah helped found Mesa Analytics & Computing, Inc., where, as chief science officer, she acts as both a consultant in the areas of drug design and compound acquisition, and as a developer of the company's commercial chemical information software products. Norah has numerous publications in the areas of fluid simulations, chemical diversity analysis, object-relational database systems, chemical cluster analysis, and chemical education. She has been the principal investigator for Mesa's National Science Foundation SBIR grants.

List of Symbols

- **Number**

$\mathbb{N}$	The natural numbers
$\mathbb{Z}$	The integers
$\mathbb{Q}$	The rational numbers
$\mathbb{R}$	The real numbers
$\mathbb{C}$	The complex numbers

- **Algebra**

π	pi
e	The natural number
a,b,c	Scalar constants
x,y,z	Scalar variables
$\sum_{i=1}^{N}$	The sum
$\prod_{i=1}^{N}$	The product

- **Logarithms**

lg	The logarithm base 2
ln	The natural logarithm
$\log_i$	The logarithm base i

- **Number Theory**

$li(x)$	The integral logarithm
$\mathcal{F}_n$	The Farey sequence or order n

- **Statistics**

σ	The standard deviation
σ^2	The variance
$\boldsymbol{\Sigma}$	The covariance matrix
ϵ	The error

- **Probability**

$P()$	The probability
$p()$	The probability density function
$\mathbf{E}[\mathbf{x}]$	The expectation
$\mathbf{Var}[\mathbf{x}]$	The variance

- **Linear Algebra**

$\mathbf{x}$	Vectors in bold face
$\mathbf{A}$	Matrices in bold face
$\mathbf{A}^{-1}$	Matrix inverse
$\mathbf{A}^{\mathbf{T}}$	Matrix transpose
$\mathbf{A} \otimes \mathbf{A}$	Hadamard product: the element-wise product of two square matrices
$\mathbf{tr}[\mathbf{A}]$	The sum of the diagonal elements of the matrix $\mathbf{A}$, known as the trace of $\mathbf{A}$

- **Set Theory**

$\mathcal{A}$	Sets
$\mathcal{C}$	Clusters
C_i	The ith cluster

- **Asymptotics**

$\cong$	Approximately equal
O	Worst case upper bound
o	Limit bound
Θ	Tight bound
Ω	Lower bound

Foreword

Clustering is basically collecting together objects, which are *alike*, and separating them from other groups of like objects, which are *unalike*; i.e., grouping and discriminating. We do this intuitively all the time in our daily lives. Without it, every object we encountered would be new. By *recognizing critical features* which make the new object *like* something we have seen before, we can benefit from earlier encounters and be able to deal quickly with a fresh experience. Equally, we are very efficient. For instance, we recognize a group of friends and relatives, even if they are in a large crowd of people, by *selecting a subset of features*, which separate them from the population in general.

Chemists, for example, even in the days when chemistry was emerging from alchemy, were putting substances into classes such as "metals." This "metal" class contained things such as iron, copper, silver, and gold, but also mercury which, even though was liquid, still had enough properties in common with the other members of its class to be included. In other words, scientists were grouping together things that were related or similar, but not necessarily identical, and separating them from the "non-metals." As Hartigan [65] has pointed out, today's biologist would be reasonably happy with the way Aristotle classified animals and plants, and would probably use only a slightly modified version of a Linnaean-type hierarchy: kingdom, phylum, class, order, family, genus, and species.

So, if we are so good at clustering and the related processes of classifying and discrimination in our daily lives, why do we need a book to tell us how to do this for chemistry and biology, which are, after all, fundamentally classification sciences? The answer to this question is twofold.

Firstly, the datasets we are dealing with are very large and have high dimensionality, so we cannot easily find patterns in data. Consider for example one of the best known datasets in chemistry, the Periodic Table of Elements. It is "exactly what it says on the tin." Mendeleyev *ordered* the 60 or 70 known elements by atomic weight, that, when set out in a table, grouped the elements by properties. So the alkali metals all came in one row in his original version, the halogens in another, and so on. He had enough belief in his model to leave gaps for yet undiscovered elements and correct atomic weights of elements that did not fit. What he did *not* do was try and cluster the properties of the elements. Even for such a small dataset, questions would arise as to how to combine, say, a density value, which is real, with a valence value, which is an integer. The datasets we are considering today very rarely, if ever have such a dominant property.

Secondly, with advances in computer power and algorithm development, analyses can be data driven, where the distinct and varied properties of the data can be taken into consideration. In the past we have carried out many analyses because we could gloss over these distinctions with a single type, rather than because we should. Data had to be regarded as real distributed in a Cartesian space so we could make use of the computing tools that were available. This is not meant as a criticism of what was done; often the outcome was useful and allowed science to progress. An example of this comes from the work of Sokal and Sneath, the fathers of numerical taxonomy [127]. They struggled with what they believed to be an intractable problem: how should one weight the different attributes of the organisms? They realized and assumed that the attributes should have equal weights because their apparent importance arose from their mutual correlations. Sneath published an illustration of this with data from *Chromobacterium*, one of the earliest papers to use computers in taxonomy [125].

In this volume, the authors present sufficient options so that the user can choose the appropriate method for their data. Again, because of the advances in the hardware and software more than one method can be used to find which is the "best." This is extremely valuable as quite often what is seen as the "best" mathematically is not the one that allows predictions to be made. So, again, the user needs to be clear about the purpose of the analysis: is it *post facto* rationalization of historical data or an attempt to make use of past data to suggest experiments to move the science forward? In the latter case, more common in drug research, higher risk and less statistical rigor may be acceptable. Having chosen clustering methods, which is appropriate for our data, we still have the thorny question of what similarity measure to use. This is governed to a certain extent by the data and the method. Our assessment of similarity varies whether we ask the question, "How alike are A and B?" or "How like A is B?" In the first form, the relationship is symmetrical. In the second it is asymmetric, i.e., more weight is given to A than B. The authors deal with this in Chapter 8. It may be that the choice can only be made by trial and error, i.e., which measure works best.

Then there are the data values themselves, which describe the objects we are trying to cluster. We are almost always choosing only a subset of the possible descriptors of an object; this may be a source of bias. If the descriptor is real valued, almost certainly it will need to be scaled. It is possible they may need weighting as well.

Only when all these things have been considered are we in the position to start our analyses; so it is clear that clustering is not a technique that sits under a button on a (molecular) spreadsheet. Practitioners in the pharmaceutical industry need an expert guide, which the authors of this book provide, to extract the most information from their data. Those of us who learned clustering from Anderberg [1], Sokal and Sneath [126], and Willett [149] now have a valuable additional resource suitable for the 21st century.

John Bradshaw
Barley, Hertfordshire

Chapter 1

Introduction

Simply put, cluster analysis is the study of methods for grouping data quantitatively. Another term used early on that nicely captures the essence of this process is performing a *numerical taxonomy*. This follows a natural human tendency to group things, to create classes whether or not these classes have much meaning. For example, when we look at the night sky we observe groups of stars, where many of the most prominent groups - *constellations* - have been given names by various cultures throughout antiquity. These appear as groups as seen on the celestial sphere, that dome of the sky above our heads. But we know now that this is a projection of deep space onto the surface of the so called sphere, and the stars within a group may differ vastly in terms of their respective distance to the earth. The groupings we see are an artifact of our perception of that projection: the *Big Dipper* is just a geometric collection of stars in the night sky, forming a geometric set of points that, by happenstance, looks to us (or looked to our forebears) like a large ladle, sauce pan, or a bear. Our present knowledge comes from new and powerful technological ways of collecting and interpreting data about stars. We now observe points of light at much greater distances by various types of telescopes, and thus distinguish stars and galaxies. Thus, we can thereby group stars into galaxies or globular clusters, or, indeed, group galaxies into a taxonomy, with classes such as elliptical and spiral galaxies, and so on, based on visual, quantitative, or derived physical properties. The spacial grouping of very large sets, galaxies throughout the universe, can also be determined quantitatively. Such grouping with large data sets is performed with cluster analysis, and the nature of such analysis is used in understanding the formation of the universe in the study of cosmology.

Clearly, quantitative tools have increased our ability to group far larger data sets than the the scores of brightly shinning stars that we see in the night sky by the use of our inherent visual perception. As more and more data was collected across the entire panoply of social and physical sciences, clustering methods were developed—somewhat independently in various fields—and found to be increasingly useful and important. With very small data sets clustering methods can be done by *hand* calculations, and often were, but larger data sets clearly require computers, and the methods now are now instantiated in algorithms and computer programs. Thus, we can now group large data sets containing protein or DNA sequence data, gene expression data, and small molecule drug-like compounds within the universe of drug discovery.

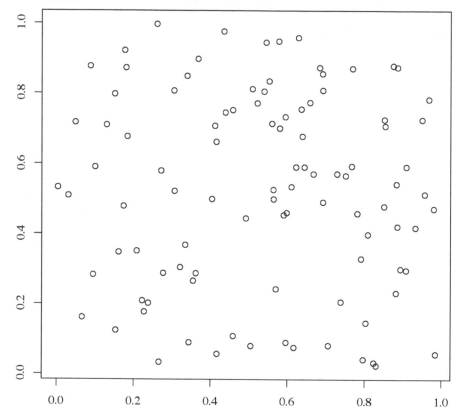

FIGURE 1.1: 100 pseudo random points on the unit square.

A nagging doubt remains however: are the groups that we find quantitatively meaningful or useful? A data set may have faint structure that may be simply indicative of a purely random process, and this is an important point throughout the study of cluster analysis, namely, one of validity. Is there strong reason to believe that the data show meaningful structure? and at what scale (e.g., local galaxies or the entire universe)? Can we make conjectures and develop hypotheses about possible classes within the domain of study? Or, are we simply grouping randomly produced structure? Take for example, the scatter of rain drops as they first appear on a flag stone or square slab of concrete in the open air: if we assume that the pattern as it develops is random, we nevertheless will see regions where the density of drops is slightly greater, and other regions where it is slightly less dense overall. We can infer nothing about these slight groupings of drops on the stone or pavement: at any point in time their pattern is just an artifact of their being a *sample* of a uniform random process. Their uniformity is at the limit: somewhat metaphorically, when the slab is completely wet. Similarly, a statistically robust pseudo random number generator on a computer will simulate sampling from a uniform random distribution, and if these numbers are plotted on a unit square as x,y pairs, they will behave very much like the rain drops: a very slight density of points will appear as in Figure 1.1. Quantitative clustering methods will group these points very much like our perception of the raindrops, even if we vary the scale somewhat at which we group the points. It is however possible to generate points with very little accumulations of density with what are known as quasi-random sequences [84].

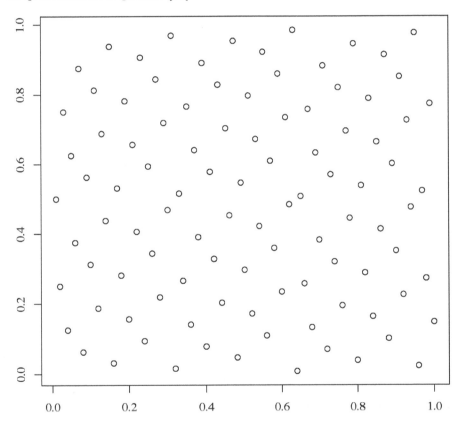

FIGURE 1.2: 100 quasi-random points on the unit square.

Figure 1.2 shows a set of quasi-random points generated from such a quasi-random sequence as the Hammersley sequence [60]. The image in Figure 1.2 is what one might naively expect a uniform random set of one hundred raindrops to look like, but such patterning is far more rare in nature than the analogues of what is found in Figure 1.1. Most clustering algorithms will indeed proceed to cluster both these types of data sets, though both results will be largely meaningless, other than to reveal that the groups formed are from a sampling of a uniform random process or a set of points generated from a quasi-random sequence. If for example this was the case with in clustering galaxies in the universe, this would certainly have consequences for the cosmology, but the groups formed thereby would not. The groups are arbitrary and they don't form classes per se.

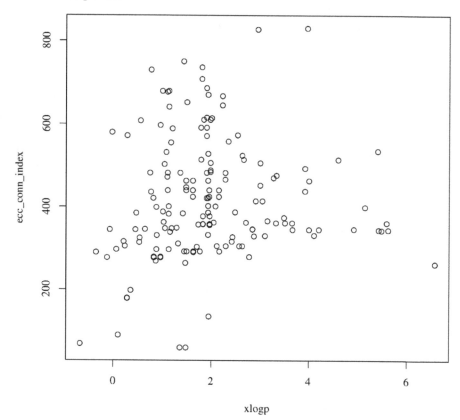

FIGURE 1.3: 161 benzodiazepine compounds plotted with respect to their eccentricity-connectivity index (a molecular graph property) and *xlogp*, a measure of a compound's lipophilicity.

On the other hand, Figure 1.3 shows real data, where groupings may or may not have meaning, depending on the context. The data in this figure are 161 benzodiazepine compounds, none with rotatable bonds, plotted in just two variables, a combined eccentricity and connectivity graph index, and the chemical property, lipophilicity. This is an artificial data set, chosen to show a simple example of what a great deal of real data often looks like; namely, amorphous and structurally confusing, at least at first glance. It is also just two properties, where in practice there are typically hundreds of not necessarily independent properties that we might consider and draw a larger subset from to perform structural and property analysis. There are dense groupings, but also a number of outliers and singletons. And again, on what scale might the groupings have any meaning is data dependent. Imagine now these same examples with many more dimensions. Figure 1.4 shows this expanded data set to include those compounds with rotatable bonds, and with ten properties in a pairwise plot. The variables were chosen to have some relation to solubility, but they are otherwise among the many hundreds of such chemical properties often used in predictive modeling QSAR studies. Some of the properties are continuous in nature such as molecular weight ("molwt"), and others are discrete, such as rotatable bonds ("rot"). It is very difficult, if indeed possible, to imagine how the data might group given these

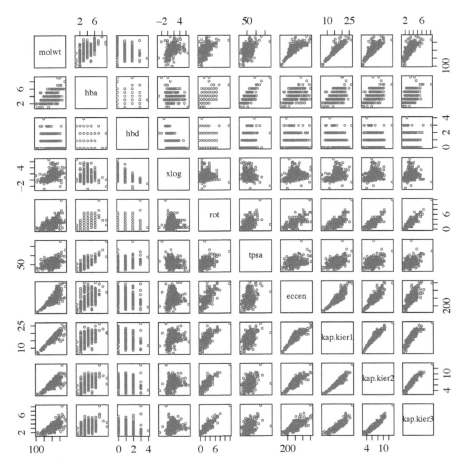

FIGURE 1.4: Pairwise plots of properties of 405 benzodiazepine compounds with mixed data types.

10 variables, or any subset of four or more of these variables. One could make an informed guess that grouping the data may be amorphous at best, by inspecting each pairwise plot and observing that aside from some outliers the data either associate in one a large group, and, in some cases, they are also highly correlated. If a researcher was trying to determine the solubility of a data set without empirical means, this is where they might start, by exploring derived solubility properties and seeing if there is structure in the data.

In contrast, the data in Figure 1.5 is nicely grouped at a certain scale into four clusters. This set is contrived and generated by random normal points in the unit square around four different centers, with standard deviations that vary in both x and y. Without this plot that shows the scale of the groups, it might take a number of tries to find that the best number of clusters for this data are four. Rarely are data so nicely grouped—and in just two dimensions

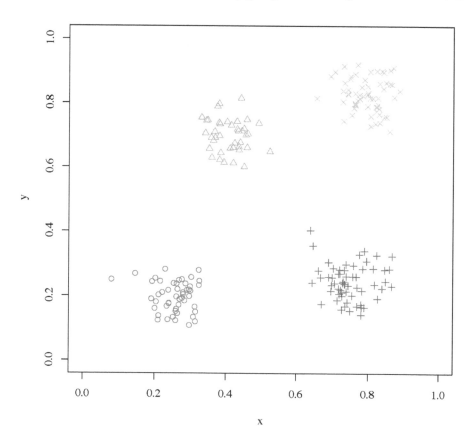

FIGURE 1.5: A contrived data set with four clusters, generated with random normal points with four centers and varying standard deviations about x and y.

that we can easily visualize. Such data that are so grouped, and are generated with processes where the distributions are known, can often be clustered using parametric methods such as mixture models. Such models would return a set of parameters such as the mean and variance estimates of the different groups—if it was known before hand that there were likely four groups, which would hopefully mean we knew something about the processes that generated the data at the outset. Figure 1.6 has the same classes as in Figure 1.5, but now with increased variance such that the classes overlap. Grouping this data will naturally misclassify the classes in the absence of other data features. Mixture models may well be the best way to group these data, given the simple generation of the classes and their distribution, but one can easily see how if the class distributions are unknown, with an increase of noise, and addition of other features, other non-parametric grouping methods may work as well or in some cases better.

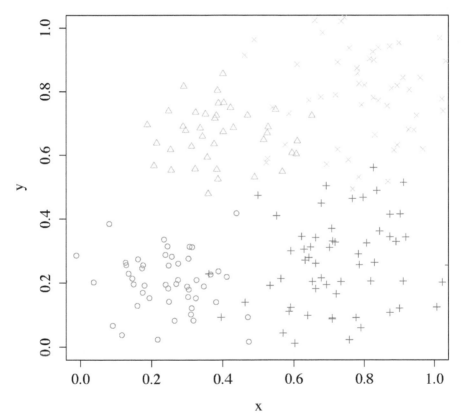

FIGURE 1.6: A contrived data set with four classes with increased noise, generated with random normal points with four centers and varying standard deviations about x and y.

A famous example of the problem of scale and random process versus class structure appeared in the 1960s and 1970s, popularly known as *cancer clusters* [101]. In health care demographics it was discovered that on occasion there were slight densities of cancer patients in a locale above the average density for a population. (For an example of cancer clusters in the workplace see [118].) The question became, were these indeed groups that were formed by some process other than random density fluctuations? Was there something in the water, in the air, in the workplaces, etc. that caused the greater incidence of a form of cancer or forms of cancers in a locale, or did these groups appear randomly? Analogously, were the data like Figure 1.2? or more like Figure 1.5? These analyses maybe confounded by other random factors in what sample of data are collected, or how the data are collected. Whether clustering galaxies, cancer incidence, gene or protein expression, sequence analysis data, or drug-like compounds, it is clearly important to consider that purely random processes and the form of the sampling can play a strong role in how groups form.

1.1　History

The early Greek philosophers had a charming habit of explaining the world around them as a set of properties, whose cardinality was often 3 or 4, or 12; e.g., there are four basic elements, earth, water, fire, and air. They posited no particular algorithms or analysis to quantify how they arrived at these groupings or classes. That is not to disparage the great intellects from the Pre-Socratics onward; they were after all using what they had at hand which were largely qualitative observations of their surroundings. And this qualitative approach continued largely unabated from Aristotle to Kant, such that many similar groupings based on qualitative observation were offered up as important classes in philosophy and science. In hindsight they seem somewhat arbitrary and without much in the way of foundation today, but nevertheless, they were well reasoned and analyzed, though nowhere are there proofs of validity from a quantitative or statistical sense. Collecting data and applying mathematics to it was relatively rare throughout the history of civilization well into the second millenium. Thus, even as the Greeks had developed extraordinary mathematics for their day, the development of mathematics and its application proceeded in fits and starts thereafter.

But, as science developed into a more quantitative discipline, efforts were made to focus on more narrow sets of observations, rather than attempting to explain all of, say, cosmology or physics with a few classes and processes. The change from (now depreciated term) *natural philosophy* to science as a more quantitatively driven endeavor, more dependent on the rigor of quantitative mathematics, has only been accelerated by computing machines and computing theory, especially now, where technologies are developed as a direct or indirect consequence of the analysis of very large amounts of data, whether it is in the development of new aircraft or new drugs. It is sometimes best to see the long historic process as a continuum, or a fine-grained incrementalism, rather than a set of large paradigm shifts or to indulge too much in the cult of personality and regard the broad changes as a result of a few great individuals. Today's scientific landscape is the result of many individuals in many disciplines working in concert in some fashion over a very long time. The development of cluster analysis is no different.

Probability and statistics were developed in the late 19th and early 20th centuries in parallel with the great advances in other sciences during this period. Exploratory data analysis and predictive modeling began to be developed to understand the increasing volumes of empirical data being collected. The advent of computing and large databases in the 1950s and 1960s provided the resources to develop more sophisticated algorithms, where larger amounts of data could be analyzed. Later, within statistics, algorithms were designed for large scale multivariate analysis, supervised and unsupervised learning, and more generally, data mining, and knowledge discovery. Today these analysis methods, of which cluster analysis is but one tool among a host of exploratory and predictive modeling methods, span science in the broadest terms - the so called hard sciences, social sciences, management science, and so on.

Articles on specific clustering approaches such as K-means and hierarchical algorithms began appearing in the late 1950s [97, 100] and general comprehensive texts on cluster analysis began to appear in the 1970s [41, 63, 126]. As computing resources increased and algorithm development progressed, by the late 1980s newer texts appeared and relied more on computing advances [76], and even provided source code for algorithms [82]. These new texts summarized in considerable detail the advances up until this time, and added new research with examples, specifically for these texts. The *Algorithms for Clustering Data* by Jain and Dubes is particularly noteworthy for its breadth and depth of treatment. It contains a very thorough chapter on cluster validity, and though there have been subsequent advances in this area (e.g., the *Gap Statistic* [135]), the chapter still stands as a good starting point for understanding this topic. Additional general texts continue to appear as well as more expansive additions of earlier texts [42]. Examples in all of these texts typically come from many disciplines or are composed of purely simulated data. Good introductory texts in statistical learning theory and pattern recognition also have concise and thoughtful chapters on cluster analysis [36, 68]. Indeed, in a review of general texts, if there is any bias towards any one discipline from which examples are drawn, it may well be image or signal pattern recognition. Some of the common nomenclature is thus drawn from this field, such as *pattern matrices* or *feature space*.

A great deal of work also was done on how to measure the similarity between data elements [54]. Literature about similarity measures continues apace, some, now so convoluted and specific to a data type and application, that many fall outside the domain of this book [46, 61, 72, 143].

New clustering algorithms continue to appear in the literature across many domains, often merely modifications of already well known algorithms to suit a specific application. One can design new clustering algorithms, but it is often prudent to thoroughly review the literature, and not just within the broad application domain—there is a strong likelihood that a new algorithm or one nearly identical to it has already been published.

Cluster analysis as an exploratory data analysis tool is dependent on the ability to visually display results and thereby help interpret the results. Thus, the visualization of clustering results has received a boost in recent years with advances in high end computer graphics, but there is still a great deal of work to be done to effectively visualize the clustering results of large data sets or data sets with large dimensions. Multivariate analysis contributes important methods aligned with cluster analysis such as principal component analysis (PCA) [80] and various forms of multi-dimesional scaling (MDS) techniques, useful for both exploratory data analysis, and data and cluster visualization [16, 31]. PCA has also been used to develop clustering algorithms for gene expression data [154]. More recently, similar methods such as non-negative matrix factorization [34] have been developed and used for both exploratory analysis and modified for cluster analysis. These tools have their drawbacks as well however, and often a combination of visualization techniques are used to help visualize the results. Hierarchies in the form of *dendrograms* can now be represented in a number of different ways and are often shown in conjunction with *heatmaps*, helping to reveal the structure relationships in the data.

Cluster analysis in bioinformatics has three common applications: gene expression data, sequence analysis, and phylogenetic tree construction. The development of gene expression screening processes led to the development of cluster analysis for the interpretation of those data. Two process forms were developed in the 1990s that test the expression of thousands of genes at once on microarrays, either cDNA or oligonucleotide arrays. Regardless of the form of the arrays used, or of the types of experiments, the general form of the data are a matrix, where the rows represent the genes and the columns represent the experiments. In some instances the experiments represented time series data.

$$
\textbf{Genes} \begin{pmatrix}
x_{1,1} & x_{1,2} & \cdots & x_{1,m} \\
x_{2,1} & x_{2,2} & \cdots & x_{2,m} \\
\vdots & \vdots & \ddots & \vdots \\
x_{n,1} & x_{m,2} & \cdots & x_{n,m}
\end{pmatrix}
$$

Expression data are typically in the thousands, and the experiments are in the tens, and rarely more than one hundred. Thus, in the matrix above, $n \gg m$. Larger expression data sets are now possible in the tens of thousands and may soon be in the hundreds of thousands to near one million. The first such expression screening results were clustered by Eisen, et al. [37] with a slightly modified hierarchical clustering algorithm that became a standard. As the size of the expression-experiment matrices change, some of the current methods may need to change on issues of size alone. More broadly, clustering approaches to various forms of biological data was surveyed in *Numerical Ecology* [92].

Broadly, cluster analysis of these data falls into three types: clustering of the gene expression data across experiments (sometimes referred to as samples; e.g., expression in response to various forms of cancer), clustering experiments across the expression data; and what is termed bi-clustering, finding distinct or overlapping sub-matrices of interest at some quantitative threshold within the expression-experiment matrix. There are numerous reasons to cluster the expression data: generating hypotheses about groups of co-regulating genes and groups that could possibly identify functional categories; genes that express in a common temporal class; and possibly determine process artifacts and error in the arrays. Clustering of the columns of experiments are often used to generate hypotheses about cell types or cancer classes. The motivation for bi-clustering takes into account that a subset of genes and and a subset of experiments (e.g., cellular processes) are of interest. Finding possibly overlapping block matrices is a hard problem, considering that it depends upon permutations of both the rows and columns of the matrix to align the sub-blocks of interest. Thus, there are a number of heuristics employed to find solutions that are hoped to be near optimal. Figure 1.7 shows a made up example of biclustered groups, colored and hatched to denote overlapping clusters.

Sequence analysis data are often clustered to find relationships or classes among protein sequences. Many of these results are then in turn used to create phylogenetic trees. Phylogenetic tree construction can however include other features and data types, but there is a common overarching connection, and that is understanding the relationships of ancestry.

Early papers on the use of cluster analysis in drug design largely concerned the structure activity relationships among small drug-like molecules [20], or determining a diverse set of small drug-like molecules [102, 120]. Hundreds of papers have since been published that involve some form of cluster analysis or similarity measures for the use therein. Almost all of the methods described in this book have been used in computational tools for the drug discovery process. The specific disciplines of bioinformatics and cheminformatics have a set of popular cluster analysis methods, related by in large to the type of data and precidence: what algorithms were first used with the new data often determined a preference for the use of that algorithm for subsequent studies. It then takes some time for researchers to try other methods, but eventually the literature includes these efforts, and in rare instances, starts a new fashion that may or may not last over time, sometimes independent of the apparent efficacy! Indeed, whether or not a particular method (self-organizing maps is a good example) is in or out of vogue in either discipline at the time of

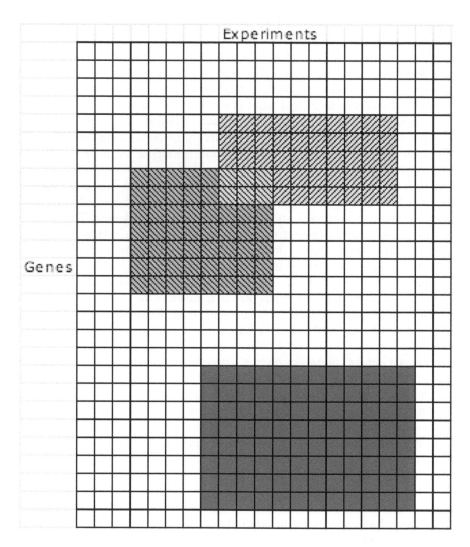

FIGURE 1.7: The sub-groups of genes and experiments align such that clusters are found in block matrices, shown here as colored and hatched to denote overlapping clusters.

the reading of this book would be an amusing time-capsule in reverse for the authors: in short, history does have a tendency to recapitulate itself, but luckily, things do change.

1.2 Bioinformatics and Drug Discovery

A large component of drug discovery involves searching for, and subsequently modifying or designing, small molecules; namely, ligands that bind to protein targets. Small molecule ligands are most often identified through *High Throughput Screening* (**HTS**), virtual studies in computational chemistry, or databases containing prior ligand activity data for a given target. Searching for drugs, or *lead discovery*, and modifying and designing drugs has traditionally been the domain of medicinal chemistry and computational chemists. The analysis and storage of this data has come be known as cheminformatics. Targets are typically identified by biologists, geneticists, and bioinformaticians, often via systems biology. However, within drug discovery the study of drug design incorporates the study of the ligand-target interaction. Thus, there are certain areas of intersection between bioinformatics, cheminformatics, and even medicinal informatics within the domain of drug discovery. Figure 1.8 loosely shows these disciplines and the important aspects of their intersection, such as genetic variation, drug design, and drug-like compound relationships. In Figure 1.9a this intersection is revealed in more detail under the common heading of the Drug Discovery Pipeline. The beginnings of drug discovery require the identification of a target for a particular disease of interest. The

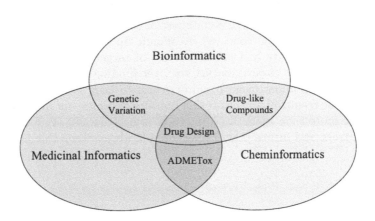

FIGURE 1.8: A broad overview of the intersection of cheminformatics, bioinformatics, and medicinal informatics within drug discovery.

identification and validation of the target requires biologists and bioinformaticians. Once the target is identified and validated the quest for finding a ligand which binds to the target requires medicinal chemists, biological screening experts, computational chemists. Once a lead series or several lead series are identified, the leads require an iterative optimization process. Again some of the same chemists, biochemists, biologists, pharmachologists, crystallographers and toxicologists all participate in optimizing the potency of a lead without introducing toxicity or other undesired effects. Successful leads are transformed into new drug entities and animal and human studies follow. Ultimately if the process is successful the resulting drug is approved for sale to the targeted market. Medicine and medical informatics is utilized in the later stages of the drug discovery process, when the drugs are administered to patient subjects.

In isolation, drug discovery topics in cheminformatics often lack the medicinal or larger biological context. The study of how ligands bind to protein targets, namely 3D drug design, is intimately connected to the study of proteins and to protein structure, and thus to the biology of drug discovery and bioinformatics. Protein and gene expression, and even to some extent phylogenetic studies [85] have become more important within drug discovery, given the area of research known as *personalized medicine* [6], whereby drugs are designed for specific groups of individuals given their genetic similarity, rather for the population at large. Bioinformatics and medicinal informatics help to pinpoint genetic variation, such that genetic subgroups that express a particular disease that is genetically related can be identified and targeted for specific drugs. Sometimes a drug developed for the general population impacts a small genetically related subgroup adversely. Having the ability to identify such persons and have them take other medication or design a new drug or set of drugs specifically for them is another form of personalized medicine that draws on the three informatic fields shown in Figure 1.9a.

Drug discovery is by in large a commercial enterprise and therefore is not always driven by altruistic concerns. Academic and government researchers have opportunities to study a broader range of targets, those which do not necessarily have a bias toward market size, yet funding sources can influence the kinds of studies undertaken. It is important to keep in mind the motivations behind the results of any clustering literature coming out of these venues if claims are more than modest. No method is best necessarily for any class of data. Current and future practitioners should always keep this in mind.

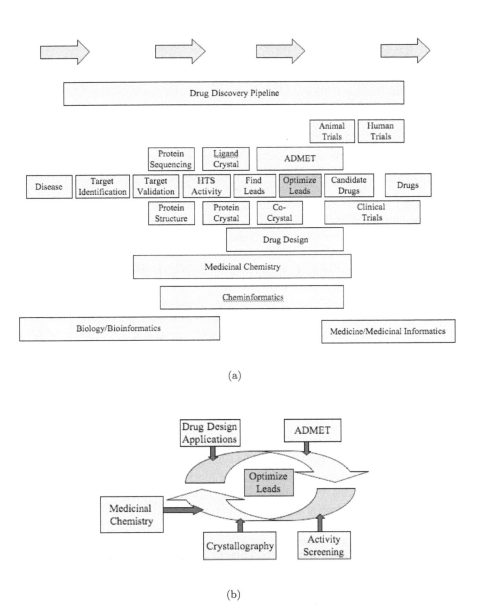

(a)

(b)

FIGURE 1.9: A schematic of the Drug Discovery Pipeline is displayed in Figure 1.9a. The iterative process of lead optimization is extracted and the necessary components drawn as inputs in Figure 1.9b.

1.3 Statistical Learning Theory and Exploratory Data Analysis

Epistemology is the philosophical study of how humans obtain knowledge. Statistics is one such tool we use to obtain knowledge: it is a quantitative study of data, and, as such, we can think of it as a window into realms of the world around us that we might not otherwise comprehend. It is a form of learning via inference and obtaining knowledge thereby. The hope is that the knowledge is somehow useful or helpful, and that it can be acted upon reliably in the future. There are two outstanding ways in which statisticians think about inference. They are referred to as *Bayesian* and *Frequentist* approaches. Bayesian concept of probability relates to measuring the state or states of knowledge. This seemingly subjective notion is formalized by methods that treat prior knowledge of events and updates probabilities with new knowledge of events. Frequentists on the other hand treat probability as the frequency of events in a system that has a definite *population* of events. This is a seemingly more objective point of view, in spite of the fact that a specific population may be quite vague or even largely unknown. Though philosophical underpinnings may differ between Bayesian or Frequentist camps concerning matters of statistical inference, each lead to the practical application of methods and algorithms designed to learn from the data.

Statistical learning theory is a portion of statistics that can be divided into two basic subheadings, *unsupervised learning* and *supervised learning*. The former concerns learning of groups or classes quantitatively—primarily cluster analysis; the latter concerns using known classes (classification) or known continuous responses (regression) to predict future cases (predict classes or continuous responses). The application of quantitative predictive methods with data sets is often referred to as *predictive modeling*. Occasionally, cluster analysis is used in conjunction with predictive modeling, but otherwise, these two forms of learning theory are distinct and their methods for validation differ. In cases where clustering is used for predictive purposes, generally it is used to segment the data to form classes, and then predictive models for those classes are developed in turn. However, these models should be regarded with considerable caution as the *classes*, so called, are of a hypothetical nature, and the noise and ambiguity inherent in the clustering derived classes are carried forward into the predictive modeling. It is not necessarily a bad idea - it may be a method of last resort, or a prescription for predictive modeling as a form of exploratory data analysis.

It should be noted at the outset to help eliminate any confusion, that the word *classification* is overloaded and has several meanings in statistical learning theory. In supervised learning theory, *classification* refers to the building of predictive models that involve known classes. However, in unsupervised learning *classification* refers to the to the basic task of assigning classes; namely, grouping or clustering items. For instance, the *Classification Society of America* is largely devoted to the study of the theory and mathematics of cluster analysis and not in general the elements of predictive classification modeling.

1.4 Clustering Algorithms

An algorithm is a step by step process that typically takes some input and arrives at a result. It can be thought of as the instructions of a calculating machine, however complicated and convoluted. The steps may be sequential or in parallel. Clustering algorithms take the data as input and return some form of a *partition* of the data into groups or a hierarchy representing some or all of the relationships among the data, from which a partition of the data can be derived by cutting the hierarchy at a hopefully meaningful level.

The algorithms can vary from these two general models such that the partitions and indeed the partitions of the hierarchies may not necessarily be disjoint—overlapping clusters are output by these algorithms. Rather than a strict partition, overlapping clusters are, to use a term from set theory, a *cover*. Fuzzy clustering [9, 157, 155] algorithms operate on data items that have a probability or coefficient of membership within clusters, and have the flavor of overlapping clusters, but are indeed distinct from these algorithms. There are both partitional and hierarchical forms of fuzzy clustering. Not all data admit to symmetrical similarity relationships. These data have asymmetrical measures of similarity for which there are asymmetric clustering algorithms of both partitional and hierarchical forms. Hybrids that contain elements from several different forms of clustering algorithms are often designed that somewhat confound the simple taxonomy of clustering algorithms outlined above.

Luckily, in general clustering algorithms and their related data structures are not particularly complicated, but as will become apparent there are typically many variations on a theme. However, in some instances there may be ambiguity as to the group membership, caused either by some decision within the clustering algorithm or in the discrete nature of the similarity measure used. Modifying the algorithm to resolve or report cluster ambiguity can complicate the implementation of the algorithm and the analysis of cluster validity.

Effectiveness often is data dependent as to what form of algorithms are applied to a specific problem. Computationally intensive algorithms are not particularly useful with very large data sets: generating a hierarchy for a very large data set may simply not be practical within the computational resources at hand, even if it might be what is desired, or prove to be the most revealing if it were doable. Whereas, the structure of a small data set may be revealed in all of its topological glory via a hierarchical algorithm, even though the algorithm is computationally expensive. The dimension of the data will almost certainly play a role as well in determining which algorithms to try. If the similarity of data items with a great many variables is very expensive to compute, it may be impractical to attempt some algorithms whose additional complexity taxes the computational resources.

1.5   Computational Complexity

Calculation takes time. Fortunately, computers are now very fast, and they will likely get faster and faster for the foreseeable future. Likewise, the size of data storage, whether on disk or in the computer memory, has increased along with computer processor speeds. This allows users the ability to compute algorithms more quickly, and do so on larger and larger problems, and store the results more efficiently and effectively. Nevertheless, it is still necessary to understand the computational complexity of a problem and to have metrics that can rank problems, and at least implicitly rank the performance of implementations of algorithmic solutions to those problems. The overall algorithmic performance should also be somewhat independent of any specific instance of the respective problem.

Some problems are not computable, some are computable but impractical (more technically, these problems are called *intractable* problems) even on today's or tomorrow's machines; some are computable, but algorithms designed to compute them may or may not be the most efficient, or efficient within reason of the resources at hand. There are some problems that are intractable, but nevertheless there are practical solutions, either approximation algorithms or *heuristics*, that arrive at solutions - solutions that are hopefully near optimal but may not be. For some intractable problems there are approximation algorithms such that solutions resulting therefrom can get arbitrarily close to the optimal solution, and thus approximate the optimal solution. They differ from heuristics where there is no guarantee of getting near an optimal solution. With a heuristic, a solution can be far from optimal. Clustering algorithms are really a set of heuristics that generate what are hoped to be near optimal solutions: namely, finding the optimal partition of a set of objects in the respective feature space is an intractable problem, where heuristics are common and efficient approximation algorithms are more difficult to design.

To understand both the time and space requirements of different algorithms in general and in terms of the size of the problem, we need to assume across the board that simple, basic arithmetic or comparison operations take roughly the same time, knowing of course in practice they don't; and that unlimited space can be randomly accessed more or less in constant time, again, knowing that in practice this is also not the case. The level playing field of these assumptions form the basis of what is known as the random access model (RAM), used for understanding the computational complexity of algorithms. In this model, constants wash away and relative algorithm performance becomes a comparison of terms purely of the problem size. This is true of both time and space requirements.

For example, sorting a small list of jumbled numbers into a ranked linear order can be done by hand, but if the list has millions of numbers, and we want to design a computer program that sorts many such lists, large and small, we will want to design an algorithm that we can program into the computer that is as efficient as possible. It turns out there are an infinite number of inefficient ways to sort a list of numbers, but only a small, finite number of general ways to efficiently do so. It has been proved in fact that for a general list of N numbers, that there is a best efficiency in the worst case, that no algorithm can be designed that can be more efficient for sorting the numbers than a certain lower bound in the costliest operation of the algorithm, that of comparing two numbers. That bound is $N \lg N$ comparisons for a general list of N numbers, jumbled in a a worst case order for sorting. Common sorting algorithms (*quicksort, mergesort, insertion sort, bubble sort*, etc.) typically have performance that varies between $N \lg N$ and N^2, and space requirements that are typically linear in N.

We can formally rank the relative complexity of algorithms with common asymptotic bounds. The focus here is when N grows large: If N is small, constants that accrue due to implementation details and other factors may overturn the relative rankings, but for most instances N need not be very large to make the relative rankings conform to the asymptotic bounds, providing hardware and software are consistent across implementations. And, if we change to new hardware or software across implementations, those rankings will remain the same.

There are three basic bounds that we will consider throughout this text. Remember in what follows that the equals sign, $=$, is really the verb, "equivalence," and not the arithmetic relation, such as $2 = 2$.

1. Bounding the time efficiency performance in terms of the worst case, what is known as the asymptotic upper bound, is denoted as:

$$f(N) = O(g(N)), \qquad (1.1)$$

where N is a natural number and there is some positive real number constant C such that

$$f(N) = Cg(N), \qquad (1.2)$$

as $N \to \infty$. This can be recast to show that there may be a non-trivial minimum positive integer, N_0, such that:

$$|f(N)| \leq C|g(N)|, \qquad (1.3)$$

as $N \to \infty$ and $N > N_0$.

2. Bounding the performance more tightly is denoted as:

$$f(N) = \Theta(g(N)), \qquad (1.4)$$

where there are two positive real number constants $C_1 < C_2$ such that,

$$C_1 g(N) \leq f(N) \leq C_2 g(N). \tag{1.5}$$

3. Bounding the performance in terms of the what is known as the lower bound, is denoted as:

$$f(N) = \Omega(g(N)). \tag{1.6}$$

In this case,

$$|f(N)| \geq C|g(N)|. \tag{1.7}$$

The Ω bound states that the performance is always more than the function $g(N)$, given some constant C multiplier. Trivially, sorting a general list of numbers is $\Omega(N)$.

These bounds provide a mechanism for understanding the performance of the clustering algorithms in terms of the size of the data. Somewhat informally, the sum of the order notations follows the following:

$$O(g(N)) + O(g(N)) = O(g(N) + g(N)) = O(g(N)). \tag{1.8}$$

Thus, for example, if a clustering algorithm first computes the pairwise similarities in $O(N^2)$, and the algorithm takes the pairwise similarities and performs an additional $O(N^2)$, the algorithm nevertheless performs all of the portions of the algorithm in $O(N^2)$ time. Only the constant C has changed. Indeed, the new C is the sum of the constants for each individual $O(N^2)$ portion of the algorithm.

For the most part simple polynomial time algorithms, polynomial in N, that find heuristic solutions can be found for clustering problems - with the understanding that the solution is very likely not the optimal partition or hierarchy that leads to the optimal partition. That is, to group N things, most clustering algorithms will take into account all pairwise similarities, of which there are $\frac{N(N-1)}{2} = O(N^2)$ such pairwise similarities in the symmetric case and $N^2 - N = O(N^2)$ in the asymmetric case. If the dimension of the data are significant, we should include that cost, in which case, the cost of computing the pairwise similarities is $O(dN^2)$, where d is the dimension. At the very least, most clustering algorithms will take the time it takes to calculate all of these similarities as a first step. (Cluster sampling algorithms that get around having to compute all $O(dN^2)$ similarities will be discussed in Chapter 5.) The time is a function of N, and in this case a polynomial in N. For polynomial time algorithms, there may be a higher exponent or more terms, or constants for each term of the polynomial, but it will be nevertheless a polynomial in N. For example, there is an algorithm sometimes used for clustering called minimum spanning tree. A common polynomial time implementation of this algorithm is $O(N^2 \lg N)$, and known fast algorithms are $O(N^2 + N \lg N)$.

Growth rates are shown for simple functions of N in Figure 1.10. Regardless of any set of constants that we multiply each function, at some point with large enough N the relative magnitude of each of these functions as they are currently displayed will be maintained. Exponential, factorial, and beyond are less efficient than any polynomial algorithm, no matter what the degree. These are intractable problems. Finding a partition of a set of N objects with d features is such a problem. There is a class of problems that, though the problem is above polynomial, its solution can be verified in polynomial time. Under the assumption of a non-deterministic solution that the problem can be verified in the polynomial time, these problems are known as problems in *NP* [30, 111, 109]. *NP* stands for non-deterministically polynomial. This terminology arises from formal automata and formal language theory. If problems can be shown to be at least as hard as problems in NP, they are known as *NP-hard*. If every problem in *NP* is reducible (known as *reductions*) to a problem p, p is known as a *NP-complete* problem, as it is both *NP-hard* and in *NP*. The consequence of this, is that if the *NP-complete* problem can be solved in polynomial time, all problems in *NP* can be solved in polynomial time. Namely, $P = NP$. Many forms of clustering problems are known to be *NP-complete* or *NP-hard*, and hence the immediate development and use of heuristics and occasional approximation algorithm to find possibly non-optimal solutions thereof. Asymptotic analysis obscures an interesting property of algorithm performance, and that is on the distribution of performance across the set of problem instances that an algorithm operates on. Consider a sorting algorithm where the a problem instance is a nearly sorted list. Some sorting algorithms may generate the solution in a time far less than a list of randomly shuffled numbers. Two algorithms may have different worst case asymptotic bounds but for some instances the more computationally expensive algorithm may be more efficient than the performance of the more efficient algorithm on those same instances. This is indeed the case with different MST algorithms, Prim's and Kruskal's MST algorithms. In addition, there may be very few instances of worst case performance, and many instances where the performance is near the lower bound, such that the broad measure of average case bound misses a broad understanding of the performance distribution over the full set of problem instances. The study of algorithm performance distribution given the set of instances is known as the empirical analysis of algorithms. MST in fact can be used for clustering data, so that understanding the types and distribution of instances for an application may be of some interest, and help to determine which MST algorithm to use.

1.5.1 Data Structures

The discussion so far has glossed over the fact that to arrive at efficient algorithms requires that the data be stored efficiently for access, and often specific storage mechanisms, data structures, are required to arrive at the best performance bounds. These are not just specific implementation details, but often integral to the order of the performance. In sorting a list for example, if we store the data in a data structure, whereby it takes a linear search in N to

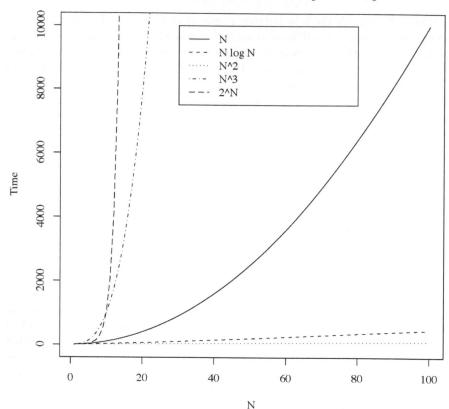

FIGURE 1.10: Relative function growth rates in terms of N and time. Shown here, linear in N, logarithmic factor $N \lg N$, polynomials N^2 and N^3, and exponential 2^N.

find the next number to consider, versus a constant time access, this will add a N factor to our sorting algorithm. Another common example that relates more specifically to cluster analysis is how a (dis)similarity matrix is stored. Say, the clustering algorithm treats the full matrix as a graph, for some graph algorithms the matrix storage adds a factor of N to the performance of the algorithm over the storage as a set of adjacency lists. From an implementation standpoint, in this case, the actual storage of the adjacency matrix will have have a constant time factor less storage than the full matrix storage. Making sure the data structure provides the best access of the algorithm problem in question is therefore a very important consideration for both time and space.

1.5.2 Parallel Algorithms

So far we have discussed algorithms in terms of single processor, and as a set of sequential processes. Some problems are fundamentally sequential, but many problems can be parallelized. In the simplest case, consider the generation of a dissimilarity matirx. The generation and storage of the matrix can be decomposed into blocks and assembled as these blocks are completed. Thus, if we have numerous processors (or processor cores), where the generation of these blocks can be distributed and then assembled in memory or on disk, we can in principle, reduce the time it takes to generate the matrix by a factor of the number of processors we use to generate the blocks, minus the time to distribute the data and then assemble it—a form of latency. Such a problem is known as *embarrassingly parallel*, meaning there is no real communication necessary among the processors other than the distribution data and the collection and assembly of the results. There are problems however, that, though still parallelizable, need to have processors communicate results or data among them. This communication cost forms another more serious form of latency, but nevertheless, it is still worthwhile to perform the parallel implementation of an algorithm over a sequential implementation in terms of the time cost. The asymptotic notation for parallel algorithms includes the order of the number of processors. So for example, a sorting algorithm may have a $O(N)$ time in the size N of the data, and with $O(\lg P)$, in P, the number of processors. It should be noted that though an problem may be parallelizable, typically the more complex the problem the less likely there is a straightforward translation of the most efficient sequential solution to a parallelizable one. One might think of this as a *solution latency* above and beyond the pre- or post-processing latency of embarrassingly parallel algorithms, or communication latency.

Increasingly, computers have multiple processors with multiple cores, allowing desktops, and even laptops the ability to perform significant parallel algorithms. Clusters of computers can now do enormous data sets in parallel. Latency may be large in practice. Say for example a you have access to 32 cores and you run your clustering algorithm on this system and it takes a day. And yet, you run the sequential version of the algorithm on a single core and it only takes 5 days. That is nothing like 32 times speed up, but if you have the resources it might be better to get the job done in a day versus waiting 5 days. But now let us say that you only have 4 cores. It may well be that the same job now takes *longer* than the sequential version on one core due to the latency. In some instances, with certain algorithms the latency is such that the speedup obtained via parallelization can't effectively grow significantly beyond a certain point *regardless* of the number of processors thrown at the problem. Thus, though the development of clustering algorithms that can be parallelized is a boon, and often independent of the nature of the latency, the amount of computing resources may be critical and will need to be carefully considered in the development and use of parallel versions of an algorithm.

1.6 Glossary

algorithm: A set of steps in sequence or in parallel to calculate some result.

asymptotic order notation: The notation for the order of computational complexity of algorithms.

bioinformatics: Biological information and analysis.

cheminformatics: Chemical information systems and analysis.

supervised learning: Classes are known.

unsupervised learning: Classes are unknown.

1.7 Exercises

1. From everyday experience list several examples of groups or classes of things. For each example, explore how the classes might be quantified (with what variables) or how they are qualitatively different. Are the classes fictitious? Are they well-defined? Are they useful in some way? are they easy or difficult to visualize as distinct groups? Now from your scientific experience and scientific data that you are familiar with, choose some examples and explore these as above. What are properties of the specific examples that distinguish the groups? Are they qualitative or quantitative in nature? Can you state how you might quantify the selection of groups, and how this might change the actual grouping? Are some groupings specious? Are some groups dangerous in that they can lead to misinformation or predictive fallacies?

2. Mechanical clustering. As late as the 1980s, determining the classification of a specific wood sample was done with a large, square, stack of cards known as a dichotomous key. All cards have a set of labeled holes along the edge of each card. The labels denoted specific properties of wood: gross properties, microscopic propertires, etc. Each card represents a specific species and the property labeled holes are either open to the edge if the property for that species is not present, and closed if that property is present. A long pin or rod that fits through the holes is used to determine each property of a new species under consideration. The pin is placed through all of the holes of one property and the cards are drawn out with the pin. The process is repeated for each property on the successive smaller and smaller subset of cards. The stack of cards is

a physical database and the pin and property holes a single query into the database of classes. The system has an implicit hierarchy of classes.

Now say, 60 of 100 properties can be determined for a wood sample. The successive use of the pin would draw out a final subset of the cards that would place the wood sample in a class of species or genus. Though the class species of each card is known, is the subset of cards together necessarily a known group of or classification? Could a new species not in the deck be determined? If so under what conditions? Explore the notion of a dicotomous key system for some of the examples you explored in the first exercise. Does the order of the use of the pin in the holes matter? How might the dichotomous key be encoded in a computer?

3. Do a literature search of Self-organizing Maps with either bioinformatics or drug discovery applications. What is the historical distribution of the number of articles that you find? Read a sample of the abstracts from the advent of this method through to the present day. Do you notice a difference in the respective claims? What roughly is the distribution of author association to academia, industry, or other institutions?

4. Below is example source code that can be used to generate Halton quasi random points, pseudo random points, clustering algorithms and plotting routines. These can be typed in directly or loaded from the website to the R prompt. Explore with the use of scale and plotting the behavior of the clustering on the data sets generated via random generators. Follow the example in source code below and change the number of points generated for each generator. What do you observe? Note, there are many statistical tests of random number generators and clustering indeed can show up pecularities of random number generators. The artifacts that you observe may reveal such random number properties.

```
Exercise 4 R code example
#You will need to install the randtoolbox R package at
http://cran.r-project.org/web/packages/randtoolbox/

library(rngWELL)
library(randtoolbox)
RNGkind(normal.kind="Box")
#You may need the following line if you are running from the
#command line
require(graphics)
#Clustering routines are in the R library "clust"
library(cluster)

numpoints <- 40
par(mfrow=c(1,1))
halton.points <- halton(n = numpoints, dim =2)
```

```
plot(halton.points, xlab="First Dimension",
        ylab="Second Dimension", type = "n" )
text(halton.points,as.character(1:numpoints))

#Repeat several times to see changes.
pseudo.points <- cbind(runif(numpoints), runif(numpoints))
plot(pseudo.points, xlab="First Dimension",
        ylab="Second Dimension", type = "n")
text(pseudo.points,as.character(1:numpoints))
#Cluster with Group Average Hierarchical clustering
        (the default)

haltonGAHclustering <- agnes(dist(halton.points))
plot(haltonGAHclustering, which.plots = 2,
        ylab="Group Average Euclidean Distance")

pseudoGAHclustering <- agnes(dist(pseudo.points))
plot(pseudoGAHclustering, which.plots = 2,
        ylab="Group Average Euclidean Distance")
```

5. The computational complexity of calculating a similarity matrix is very simple $O(N^2)$. However, given the dimension d, what implementation details impact the constant implied by $O()$ notation?

6. The minimum spanning tree of a graph is a very simple notion, but efficient algorithms to compute the minimum spanning tree of a graph can be quite complicated, requiring special data structures. Look up the various versions of computing the minimum spanning tree in a graph algorithm book, or on the Web and compare the computational complexity of the algorithms. How might the constants for each algorithm impact ones choice of algorithm to implement given the size of the typical problem one might address—many problems with small N, or a few problems with very large N? Can you think of how you might use the minimum spanning tree to form clusters?

Chapter 2

Data

Data in bioinformatics and drug discovery are manifold. The data are often empirically collected and input into a computer either by hand or automatically; or data can be derived from known properties and thus generated on the computer. These data are then typically placed in a database or flat file for analysis. Of course, within any given data set there may be error, missing data, erroneous elements, mixed types, or a mixture of empirical and derived elements. As a result, data quality analysis and expert knowledge about the data are almost always necessary to go forward with meaningful exploratory data analysis. Having a thorough understanding of the data, its generation, its scale, its range, its anomalies and outliers, goes a long way in determining the type of clustering algorithms to apply, and how to evaluate and validate the results. How the data are represented internally within a computer is also important: the precision, memory usage, and disk space used to represent the data can impact the choice and effectiveness of the cluster analysis performed.

In order to input data into a clustering algorithm, additional processing is often required. The data may need to be scaled, normalized, or otherwise transformed. A measure of proximity between data elements will then need to be chosen. Depending on the clustering method, the data matrix (sometimes called the *pattern matrix*) or some form of a proximity matrix can then be used as input. In turn, there are a number of different ways to consider a proximity matrix, either for matrix manipulation through matrix algebra, or as combinatorial graph objects. Each also has ramifications for how one may visualize and interpret clustering results, or design, or intuitively understand clustering algorithms.

2.1 Types

There are four basic data types, binary, ordinal (count data), continuous (ratio data), and categorical data. These are not hard and fast types in that they can be converted from one to the other, and for any particular data set, there may be several different types that make up the dimension of the data. Thus, for example, a set of small drug and drug-like molecules may have binary data that represents the chemical graph structural properties of each molecule, a collection of calculated properties such as the molecule's molecular weight, the number of rotatable bonds, a measure of lipiphilicity, and so forth. The binary data are typically a fixed length string of 1s and 0s, whereas the number of rotatable bonds is in the set of whole numbers (that has an effective range of between 0 and, say, 30). The molecular weight and measure of lipiphilicity are continuous valued data, but the precision of these values make them effectively transformable to, though large, a nevertheless finite set of whole numbers. Regardless of continuous valued nature of the numbers, some form of computer (or machine) precision, or arbitrary precision set by a programmer, defines these values in a finite scale. As long as the scale is large, the discrete nature of the computer representation behaves very much like continuous values. The set of molecules may also have categorical data, say, that each compound has a unique identifier assigned by a company or chemical vendor, for example. This data may be important in some applications, but is quite distinct from numerical data, as it has no ordering, nor any measure of similarity absent some other property.

For an application, it may be that just one or a few of these types are pertinent to the clustering application at hand. For example, one may want to group the compounds on their chemical graph structural properties alone, or on their calculated properties alone, etc. It is often convenient if the data are all of the same type. This helps to narrow the choice of similarity measure to those common to that type. In such cases, transforming to the most common type is often the typical, though not necessarily best, solution. Again, this can be very data and application dependent.

2.1.1   Binary Data

Binary data are common in drug discovery and a bit less so in bioinformatics. How a medicinal chemist can input a known compound into a computer is a question of encoding. Once a set of compounds are encoded into a machine readable form, a way of searching that set and operating on that set is within the domain of chemical information systems.

In the mid 1950s the Wiswesser Line Notation (WLN) was introduced and enjoyed widespread use as a common chemical encoding [152]. In the 1980s one of the most commonly used encodings still in use today is what is known as SMILES [145], an ASCII string containing an extended alphabet and a grammar that operates on that alphabet to define a chemical graph encoding. A SMILES encoding of benzene is simply *c1ccccc1*. A caffeine SMILES string is *c12c(n(c(n(c1=O)C)=O)C)ncn2C*. These strings are not necessarily unique, but given a set of rules, SMILES strings can be canonicalized, such that they can be made unique [146]. The OpenBabel canonicalizing program outputs the following SMILES string for caffeine, *Cn1cnc2c1c(=O)n(C)c(=O)n2C*. The order of the atoms has changed but the chemical graph is the same [3]. A more recently adopted encoding is what is known as the International Chemical Identifier (**InChI**™) string. InChI strings were created to provide an open standard for identifying chemicals in chemical information systems. A desire to utilize InChI strings in on-line database systems [134] motivated the simplification of the InChI strings into InChI keys [69, 86] the most recent version provides the desired standardization and the strings and keys are named **Standard InChI** strings and Standard InChI keys. In Figure 2.1, the Standard InChI string and key are given for caffeine. The the respective SMILES, Standard InChI string, and Standard InChI key representations of caffeine molecule in Figure 2.1 are:

- Cn1cnc2c1c(=O)n(C)c(=O)n2C

- 1S/C8H10N4O2/c1-10-4-9-6-5(10)7(13)12(3)8(14)11(6)2/h4H

- 1-3H3,RYYVLZVUVIJVGH-UHFFFAOYSA-N

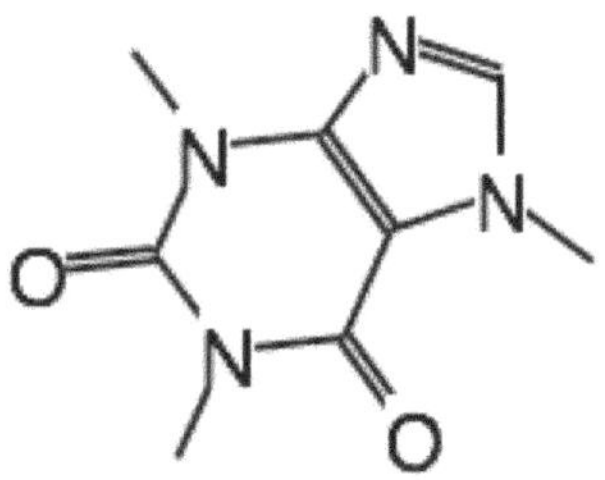

FIGURE 2.1: A 2-dimensional depiction of the caffeine molecule

Whether utilizing InChI strings or cannonical SMILES for exact structure lookup in chemical database systems, searching for substructure matches or similar compounds within a database using a string similarity measure (e.g., edit distance or n-gram analysis [57]) is somewhat inefficient. Thus, for database storage and retrieval a further encoding can be abstracted by various forms of subgraph matching that converts the subgraph properties of the chemical graph as defined by the SMILES string into a string of bits of fixed length, typically in the hundreds or thousands of bits long. Subgraphs can be defined by another language known as SMARTS [73]. In Figure 2.2 a key-based fingerprint of 1s and 0s is shown for a compound, and the specific subgraph key bit is highlighted, matching the highlighted molecular structure, a ring with a non-carbon atom. The first string at the top of the figure is the SMARTS string denoting the [*Non-carbon atom in a six-membered ring*. The image in the figure is from an interactive, dynamic web application, where one can upload a compound SMILES string, create the fingerprint and highlight the key bits, and, if the highlighted bit is "1," see the corresponding SMARTS key string and the matching subgraph(s) in the compound. Figure 2.2's key-based fingerprint depends on a pre-existing dictionary of chemical *key definitions*, typically encoded as SMARTS strings (though they need not be), that comprise the fingerprint. Binary representations can be generated from hashing of path based molecular graphs (essentially, hashing induced subgraphs, where there are unique paths of a length no greater than a specific threshold path length), keys generated from structure dictionaries or pre-defined keys based on relevance to drug discovery [91]. Originally designed for database searching, more commonly referred to as *similarity searching*, and the fingerprints were subsequently used to cluster compounds [20, 149]. Bits in common among a set of fingerprints for a mode or *modal* fingerprint [121]. A modal fingerprint can be used to determine a common substructure among all of the structures represented by the fingerprint set. The modal can be parameterized such that not all bits need be in common, but some typically high percentage of the bits need to be in common. The modal fingerprint can then be used to search through a different database of fingerprints as a substructure searching procedure, or a way of visualizing the common substructure in a set of compounds.

Similarity searching and clustering applications compound selection are motivated by what is known as the *similar property principle* [79]: namely, similar drug or drug-like compounds have similar biological activity. Whereas there is considerable evidence that this is true, it is confounded by factors such as molecular shape, conformational energy, and electrostatics. These can be included in a broader definition of *structure*, but it is difficult to encapsulate or encode all of these factors into a single, efficient, data type. So much effort has been expended in designing structural, *2D*, binary fingerprints and binary measures of similarity, trying to find the best mix that correlates structure to biological activity. These in turn have additional biases, where the size of the molecules play a role, leading researchers to navigate the choice of fingerprinter and measure best suited to their data [35, 46, 72, 147].

Many various forms of binary fingerprints have since been designed, and of various fixed lengths, though the lengths are often in multiples of 32 or 64 bit words, making them more efficient for both storage and calculating similarity. Fingerprint methods can be either general or data set specific. Early fingerprinting methods developed by software vendors, MDL, Daylight, and BCI are still in use today [73, 25, 129]. More recent additions continue to be developed, especially those of a data specific natures. Other common molecular formats are the Protein Data Bank (PDB) file, the MOL2 file, and the Structure-Data File (SDF). The fingerprints are often independent of the graph encoding such that they can be derived from any number of these graph encodings, not just SMILES strings.

The SMARTS subgraphs can be very specific (e.g., a benzene ring), or very general as in the example in Figure 2.2. Similarity measures for fixed length bit strings are very fast to compute. Though there is a loss of information, searching the database by these binary string encodings is a great deal faster, and the storage of those strings is also less in general, as they can be stored as bit per bit. There are many efficient similarity measures that can be used

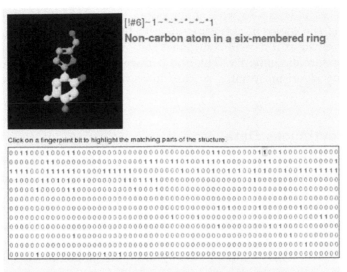

FIGURE 2.2: Example key based compound fingerprint is displayed below a molecular depiction. A 1 or *on* bit is highlighted along with the matching feature in the molecular depiction. The corresponding SMARTS key with description is displayed next to the structure Jmol image [78].

to compare two fixed length bit strings, some, however, are more effective for a particular application than others.

2.1.2   Count Data

Count data are quite common across most disciplines, and bioinformatics and cheminformatics are no different. The number of rotatable bonds is a one dimensional count variable for compound data, but there can be many dimensional count data. Consider the binary representation of chemical structure: there can be more than one distinct subgraph key within a molecule. In the binary representation, if the subgraph key is extant in the molecule, the bit is turned on. However, there may be more than one instance of that subgraph key. Say, the subgraph key is the benzene ring and there are three distinct benzene rings within the compound, a count representation would signify the number 3 for that key, rather than just turning on that bit key. In the nomenclature of cheminformatics, this is known as the count fingerprint. Such counts are often across a small range and lead to issues of comparison similarity to binary data.

Changes from type to type can alter the properties of measures used to compare them. The Tanimoto or Jaccard measure is a metric with binary data, whereas, it is not so with count data. This can have important consequences in the choice of clustering algorithms used with each. Count data are often treated as if it were continuous data and the Euclidean distance is used to measure dissimilarity. This too can lead to odd effects of ambiguity in clustering algorithm results, so care must be taken when evaluating the results.

2.1.3   Continuous Data

Continuous data are really not continuous at all when represented in a computer. However, with a sufficient range, precision, and size (large number of values), such data for the purposes of data analysis can behave like continuous values. Nevertheless, it is wise to understand how the values are represented, with what precision, their range, how they will behave under transformations, and any properties of their possible recording error. Any elementary text on numerical analysis will have a discussion on rounding error and computer arithmetic [21].

2.1.4   Categorical Data

Categorical data are a form of labels; e.g., "yellow," "red," "blue," where there is no sense of measure or distance between them. They are themselves classes. That is not to say that we can not use this data in conjunction with other data of different types and design a measure that takes into account all of the types. If the data are a response, say a class of activity (inactive or active) or expression (down regulated, up regulated), these data may be more appropriately used for supervised learning, namely, used to build predictive classifiers. If there are just two categories, these can be transformed to "1" and "0" to put the categorical data on a numerical footing.

2.1.5   Mixed Type Data

Mixed data presents significant issues in terms of the choice of transformations, the introduction of precision for binary or count data, and the choice of measures then used. Binary data may be converted to count data, where the counts are simply 1 and 0, or count data my be transformed to binary data, etc., such that all of the data are of one type. Then an appropriate measure of that type can be used to compare data items. In addition, given the different types, some care and thoughtfulness about the possible need for weighting or scaling the transformed variables should be taken.

2.2   Normalization and Scaling

Different dimensions of the data may have different ranges: they may be on completely different scales. If we were to cluster a classroom of students on their height and weight, it is clear that these two dimensions are on different scales and the data values will almost surely have different ranges. However, it is hard to know necessarily if it is wise to leave the data on their respective scales, or to place them on the same scale; and if the latter, what scaling to use. Again, this goes back to expert knowledge about the data in question. Some scalings can be detrimental to the clustering process. The best clustering separation for the data may be the raw data or some combination where some features are scaled and others are not. Figure 2.3 shows simulated data where there are clearly two well separated clusters and when scaled in one dimension (feature), this separation disappears. With small data sets with just a few features, this property can be done by simple inspection, but for any data set of significant size and dimension this becomes a much more difficult exercise.

The most common scaling is the z-value scaling, putting the values on a range determined by the units of the standard deviation of the sample, with a mean of zero. Normalization to the range of 0 and 1 can also be made which may have curious effects with outlying values. Outliers in the data can greatly impact clustering results, so normalization or scaling of the data may be helpful or detrimental in this regard. Because such changes in clustering results are so data driven there is really no set preference or preferences for normalization or scaling methods. It may be best to perform various preliminary outlier detection routines on the raw data and making decisions at that time what methods should be performed going forward, or even removing outliers if the situation warrants.

2.3   Transformations

Modeling and interpretation of data tends to be far easier if the data are on a linear scale, and in many cases non-linear data can be transformed to a linear scale. The left hand panel of Figure 2.4 displays an exponential transformation of raw feature data. The exponential transformation of the data in the right hand panel exposes some heteroscedacity (a change in variance as the values increase), which was not nearly as apparent in the data in its original form on the left. It may be hard to determine absent expert knowledge whether to transform any number of features, and it may be the case where inspection and trial and error are the only guides.

Recorded values may be on a scale such that taking the log of the data

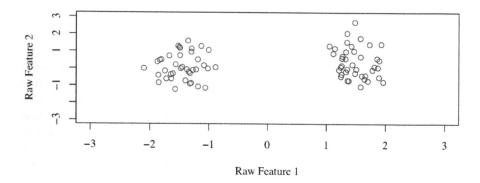

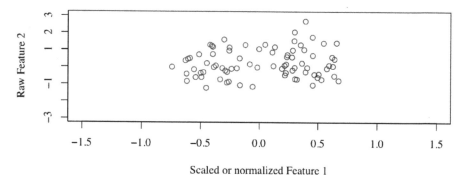

FIGURE 2.3: Artificial example of a scaling or normalization problem with clustering. Two groups clearly separated in the raw data, though once scaled or normalized along Feature 1(x-axis) are no longer clearly separate.

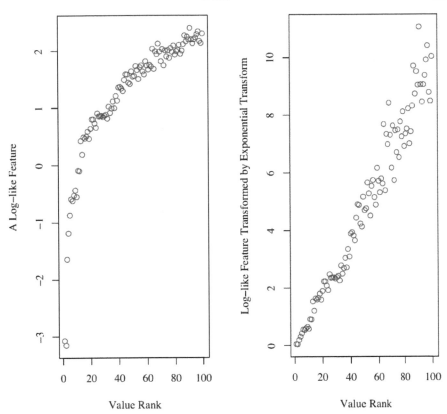

FIGURE 2.4: The raw feature data on the first plot has been ranked by value and has a log or log-like appearance. The feature is exponentially transformed in the second plot and appears linear after the transform, but heteroscedacity, barely noticed in the raw data, is revealed and magnified by the transformation.

transforms the scale to a linear scale. IC_{50}, the half maximal inhibitory concentration, is one such empirically derived set of values. IC_{50} values relate to what concentration of a compound is needed to inhibit a specific chemical or biological process. Generally, the hope is in drug discovery that it takes just a very small concentration, namely, a small dosage of a drug compound to affect the inhibition. Thus, the values can be very small—in the pico- or nano-molar concentration and range upwards by orders of magnitude. Taking the negative log of such values transforms the IC_{50} values to a positive linear scale, where the greatest positive values are the most inhibitory.

2.4 Formats

Data are the precursor to clustering. Much preprocessing and focus can be placed on the data that is to be clustered, given the many available formats. Data extracted from databases is typically in or can be easily converted to a standard usable format. Whether proprietary or public, databases in Bioinformatics and Drug Discovery in general can be an analyst's friend or foe for this reason. Flat files of data are still commonplace in the industry.

There are many publicly available gene and protein sequence databases: *GenBank, GenPept, European Molecular Biology Laboratory (EMBL), Protein Sequence Database (PIR-PSD), RCSB Protein Data Bank (PDB), NCBI Gene Expression Omnibus (GEO)* represent some of the major repositories [38, 47, 113, 112, 4]. BioCyc contains databases of metabolic pathways/genomes [81]. The larger collections of chemical structure databases are available through chemical vendors directly, their webservice providers (e.g., *eMolecules*) [39] or academic (*ZINC*) [74], journals, or government institutions, (e.g., *PubChem*)[48]. Assay data for small molecule ligand binding has recently been more available to the public (*PubChem*) [48]. Such data has been determined to be proprietary by pharmaceutical and biotechnology companies, so there is a lot of data of this type inside companies that have drug discovery markets. The list of chemical formats is quite vast. A few common formats are InChI, SMILES, SDF, Mol2, etc. [91]. Typically each drug design software vendor has its own proprietary format, but will input and export to many of the more commonly used formats to facilitate interoperability. Databases and formats change and may go in and out of vogue, but many now are quite stable and have had continued use for many years—namely, they have stood the test of time.

2.5 Data Matrices

Formats and databases aside, in the abstract, the data will be in the form of a matrix, where the rows represent the data items, and the columns represent the dimension of the data. There will be N rows and p dimensions. Informally, the extremes of data matrices for clustering will fall under the following headings: *long and tall* data are where $N >> p$, and N is large; *short and fat* data are where $N < p$, and N is small. How to treat these extremes in terms of cluster analysis typically relates to the computational complexity of N in the former case and p in the latter case. For example, there is a strong chance that *short and fat* is over determined in the sense that many of the variables may be highly correlated. Dimension reduction of p is likely called for as a first step after scaling and transformations. The choice of clustering algorithms may in turn be restricted by the computational resources implied by either the size of N or the size of the dimension, p, the number of variables, relative to N.

In some cheminformatics applications N can be very large, on the order of several millions and up, and p may also be in the hundreds, or, in the case of binary or count fingerprints, in the several thousands. Expression data dimensions are typically smaller, but can be by no means insignificant.

2.6 Measures of Similarity

Similarity and its mirror, dissimilarity, quantitatively measure how close two things or a collection of things are to one another. This is within a *space*, defined by the data type and the measure. There are many ways to quantify similarity given the data type. The Euclidean distance between points in simple geometry is one such measure of dissimilarity. The measure or distance, d, between two points $\mathbf{x}$, and $\mathbf{y}$ in Euclidean space with dimension P is

$$d = \sqrt{\sum_{i=1}^{P} (x_i - y_i)^2}. \tag{2.1}$$

Euclidean geometry maps intuitively relate to every day experience where we have a sense of near and far so it is a simple and easy measure to understand. But measures of similarity and dissimilarity apply to the type of data: not all data are conveniently transformed to plane geometry, or even to the generalization of many dimensioned geometry for that matter. Thus, for binary data there are a host of measures, some simple and common, some complex and obscure, and similarly for count and continuous data.

Much is made of whether or not a a measure is a metric: that is, does all of the data admit to the triangular inequality. We can see this most clearly with the use of the Euclidean distance, but it takes a bit more mathematics to to show that other measures are or are not metrics [31, 54]. Simply, the following properties of $d(\mathbf{x}, \mathbf{y})$ in general, for any data items under consideration $\mathbf{x}$, $\mathbf{y}$, and $\mathbf{z}$:

- $d(\mathbf{x}, \mathbf{y}) \geq 0$ (**Nonnegative**)

- $d(\mathbf{x}, \mathbf{y}) = 0$ iff $\mathbf{x} = \mathbf{y}$ (**Reflexive**)

- $d(\mathbf{x}, \mathbf{y}) = d(\mathbf{y}, \mathbf{x})$ (**Symmetric**)

- $d(\mathbf{x}, \mathbf{y}) + d(\mathbf{y}, \mathbf{z}) \geq d(\mathbf{x}, \mathbf{z})$ (**Triangle Inequality**)

The metric property is often important in what clustering algorithms can be used with various types of data. Thus, measures can be classified by type, and whether or not they are a metric. Clustering algorithms can then be classified as to what data types and measures can be used effectively. By leaving out the triangle inequality, we can define what is known as the *semi-metric* property. Again, some clustering methods operate on both metric and semi-metric measures. Hierarchical procedures can induce a metric on a more general dissimilarity measure that is not a metric. Metrics and induced metrics under hierarchical procedures form what is known as an *ultrametric* such that we can add one more property with the additional constraint:

- $d(\mathbf{x}, \mathbf{z}) \leq max(d(\mathbf{x}, \mathbf{y}), d(\mathbf{y}, \mathbf{z}))$

This latter property can be reflected in an matrix such that respective matrix indices i, j, and k of $\mathbf{x}$, $\mathbf{y}$, and $\mathbf{z}$ form the following relationship:

$$d(i, k) \leq max\left(d(i, j), d(j, k)\right) \tag{2.2}$$

where $d()$ here represents the element value defined for the row, column pair, for all i, j, and k. This matrix has the property of ultrametricity as will be discussed in the *proximity matrices* section below.

All of the measures in this section are symmetric: the similarity between any two data elements A and B is the same as the similarity between B and A. Some measures of similarity are not necessarily symmetric however. These measures are more often found in the social sciences. Take for example a social scientist who wishes to find subgroups of like-minded friends among a large group of people who all know each other. The scientist designs a measure of 1 to 10 as to how much each person likes the other and canvases them as to their preference for the other members of the group. Individual A for example may judge individual B with a score of 8, whereas individual B may score individual A with a score of 7. An N by N proximity matrix based on these preferences would likely be asymmetric. The scientist would then be faced with trying to find like-minded groups (namely clusters, and informally, *cliques*) among the entire group. There are a number of clustering methods that can be used to perform asymmetric clustering, or transformations of the proximity matrix that transform the asymmetric matrix to a symmetric matrix so that symmetric clustering algorithms can be used. These measures and methods concerning asymmetry will be discussed in further detail in Chapter 8.

2.6.1   Binary Data Measures

Measures of similarity for binary data are not particular intuitive: regardless of the measure, the space is very different from simple Euclidean geometry. Rather than thinking of the space of the binary data, a common notion to keep in mind is that of the number of bits in common among any two binary bit strings in the data. This suggests a measure of similarity and its complement, a dissimilarity or *distance*. Most binary measures have this as an underlying idea—there is less concern quantitatively of the bits turned off. If the measures use the number bits turned off and the number of bits turned on, the size of the bit strings becomes more important. Some measures include the size of the bit string explicitly. The underlying data will have a great deal of importance as to which measure or measures are used. There are two general forms of binary measures, *associations* between the binary strings, and *correlations* that are analogous to the correlation measures found for continuous valued data, such as the common Pearson correlation coefficient. Where the use of correlation is more appropriate for the data, and the size of the data are not too prohibitive, these measures can be used.

2.6.1.1 The Tanimoto or Jaccard Measure

The most common measure of similarity for binary data are known as the Tanimoto in the drug discovery literature. It is more commonly referred to as the Jaccard coefficient in other fields; a name which pre-dates the Tanimoto by many decades [75, 130]. Its compliment, $1 - Tanimoto$ is known as the Soergel dissimilarity. In the cheminformatics clustering literature, authors often refer to the measure they used as the Tanimoto, though in general it can be inferred that they used the Soergel when operating on the data for their clustering algorithms. It is a simple measure that compares what is most like among two bit strings, counting bits in common. There are two formulations. The first and somewhat simpler to understand, starts by counting the bits in common between the two strings, A and B. Let the bits in common be c. Let a be those bits on in A but not in B; and, let b be those bits on in B but not in A. Then, the Tanimoto similarity, s,between A and B is

$$s = \frac{c}{a+b+c}. \tag{2.3}$$

The range of s is of course $[0, 1]$, and it can be turned into a dissimilarity, or informally, a distance, via $d = 1.0 - s$ In the other formula, let c be all of the bits turned on in either A and B, and let a and b be defined as a equals all of the bits turned on in A and b equals all the bits turned on in B. Then the alternate formula for the Tanimoto is

$$s = \frac{c}{(a+b)-c}. \tag{2.4}$$

An equivalent formulation set notation is sometimes found in the literature:

$$s = \frac{|\bigcup(A, B)|}{|A| + |B| - |\bigcup(A, B)|}. \tag{2.5}$$

It is typically more efficient to generate the Tanimoto measures by implementing Equation 2.4, but it is often best to see all three of the equations above to quickly get a sense of the measure. It is easier to see directly the discrete nature of the measure, given Equation 2.3. Note, that if N is the length of the fixed length bit strings, the Tanimoto can generate all simple fractions (in the range of $[0, 1]$), where the numerator is less than or equal to the denominator, and the denominator is less than or equal to N. There are only so many such fractions, and when all fractions are reduced (also known as the Farey sequence N, F_N), in the average case there are only $O(\frac{3}{\pi^2} N^2)$ such fractions. This means there are not that many possible measure values, especially if N is small. The discrete nature of some data types and measures are important to keep in mind, especially since their distribution may be skewed and not at all smooth. Table 2.1 lists the possible values of the Tanimoto coefficient, given the length of the binary string in question. Since there are many binary string lengths in use, even ones that are quite short [153], and the Tanimoto measure so ubiquitous, it is good to keep in mind the number of possible values that can occur and data type representations that can be used to store them should the number of comparisons be very large. The typical data types found in say, C and C++ such as *unsigned short int* (unsigned 16 bit word), *float* (32 bit floating point precision), and *double* (64 double precision), for each set of values is identified. This shows the precision necessary to keep the values distinct for each binary string length. This topic will be discussed in greater detail in Chapter 8.

TABLE 2.1: Possible Values of the Tanimoto Coefficient (Number of Reduced Proper Fractions), Given the Length of the Binary String in Question (the Typical Data Type and Its Common Respective Bit Storage for Each Set of Values is Identified)

Binary String Length	Farey Numbers	Data Type Storage	Data Type Bit Length
64	1261	short int	16 bit word
128	5023	short int	16 bit word
256	19949	short int	16 bit word
320	31233	short int	16 bit word
512	79853	float	32 bit word
640	124539	float	32 bit word
768	179409	float	32 bit word
1024	318965	float	32 bit word
2048	1275587	double	64 bit word
4096	5100021	double	64 bit word

The Tanimoto (and the Soergel) is indeed a metric with binary data under the following transformation [31], given the earlier definitions of a and b above:

$$s = \sqrt{\frac{c}{a + b + c}}. \tag{2.6}$$

This fact means that it can be used with more clustering algorithms that depend on the metric nature of the measure used.

The Tanimoto measure has a slight bias towards making bit strings that both have a large percentage of bits turned on as being more similar than two bit strings with only a few bits on in each. This bias can have an impact on clustering results, and may bias the *centrotype* size (number of bits turned on in what might be more loosely and perhaps inaccurately referred to as the *centriod* of a cluster) relative to the other members of a cluster.

2.6.1.2 The Baroni-Urbani/Buser Coefficient

The Baroni-Urbani/Buser (BUB) measure was developed to study the probability of various binary coefficients in the context of the Jaccard measure. Like the Dice coefficient (see in Other Binary Measures) it has a similar form to the Tanimoto:

$$s = \frac{\sqrt{cd} + c}{\sqrt{cd} + a + b + c}, \tag{2.7}$$

where d is the number of zero bits in common to both binary fingerprints. Again, and for what follows a and b are the earlier definitions of bit turned on but not in common to A and B. BUB has been shown to be roughly equivalent if not marginally better than the Tanimoto with respect to structural fingerprints in similarity searching and clustering of drug or drug-like compounds. The square root term helps to remove the size bias common in other measures such as the Tanimoto and Euclidean. The d terms helps to factor in the size of the fingerprint and changes to a small degree the distributional properties of the measure versus the Tanimoto. Figure 2.5 shows a comparison between the Tanimoto and the Baroni-Urbani/Buser coefficients given all binary strings of length 5. Both are discrete and have a fractal ruler pattern, where the Baroni-Urbani/Buser tends to have more larger values. Binary strings of length 6 are used in Figure 2.6. With binary strings of length 5 there are fewer possible values given the Baroni-Urbani/Buser coefficient than that of the Tanimoto. The opposite is true of these coefficients given binary strings of length 6. This variability in the number of possible values has to do with the number of divisors of the length of the binary string. When the length of the string is a prime number the number of possible values tends to be less with the Baroni-Urbani/Buser coefficient. The distributions of the Tanimoto coefficient for a 6 bit string are displayed in Figure 2.7 and the distributions for the Baroni-Urbani/Buser coefficient for bit strings of length 6 are given in Figure 2.8. These distributions are discussed in greater length with regards to clustering ambiguity in Chapter 9.

The fractal nature of measures such as the Tanimoto and the Baroni-Urbani/Buser can be revealed by plotting the similarity matrices of all pairs of distinct fixed length binary strings as a heatmap of values. Figures 2.9 and 2.10 are the heatmaps for the 6-bit distributions for the Tanimoto and Baroni-Urbani/Buser measures respectively. Each exhibit a fractal-like appearance akin to a well known fractal pattern called a *Sierpinski Carpet*. The distributions of these measures in practice behave very much like probabilistic fractals. Modeling such distributions have subtle issues that are somewhat unusual in the context of common data analysis.

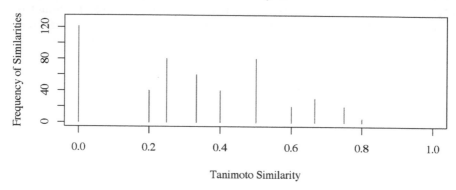

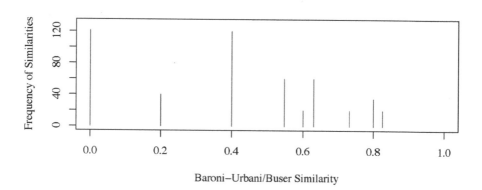

FIGURE 2.5: Comparing the Tanimoto (above) and Baroni-Urbani/Buser (below) coefficient distributions, given all binary strings of length 5. The right hand tail of the distribution of the Baroni is slightly larger and nearer 1, but it still very discrete and has fewer possible values (9 to the Tanimoto's 10).

With careful programming the Baroni-Urbani/Buser measure is roughly 40 to 50% slower than a fast implementation of the Tanimoto measure. This makes it competitive with the Tanimoto for large scale compound selection and related similarity searching and clustering applications, and hence its more recent widespread use, as it has slightly better distributional properties than the Tanimoto.

2.6.1.3 The Simple Matching Coefficient

The simple matching coefficient is just that: just count the bits turned on in common between A and B (namely c as defined for the Tanimoto measure),

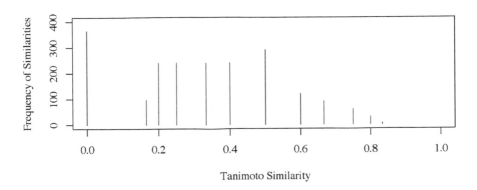

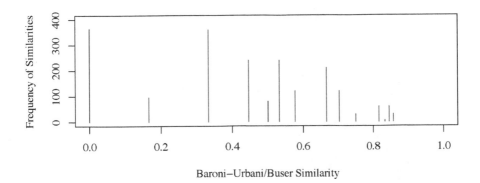

FIGURE 2.6: Comparing the Tanimoto (above) and Baroni-Urbani/Buser (below) coefficient distributions, given all binary strings of length 6. Now there are more possible values of the Baroni than the Tanimoto. This characteristic changes depending on whether the length of the binary string is a prime number or not. Both have a fractal nature.

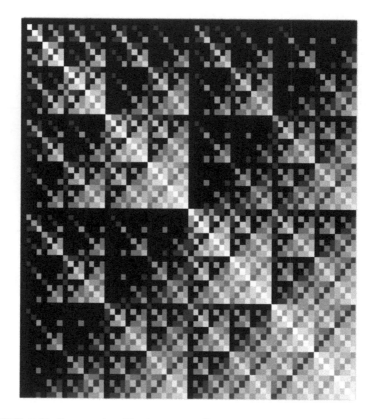

FIGURE 2.7: Comparing Tanimoto coefficient distributions, 6 bit per string. Black corresponds to no color or 0 and white corresponds to maximum color or 1.

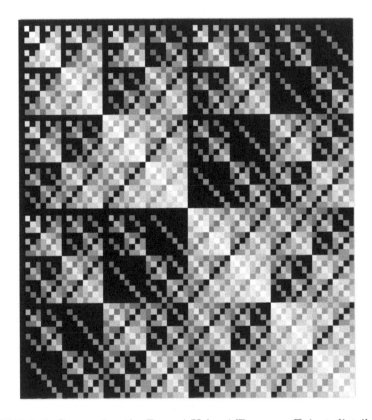

FIGURE 2.8: Comparing the Baroni-Urbani/Buser coefficient distributions, 6 bit per string. Black corresponds to no color or 0 and white corresponds to maximum color or 1.

and let d represent the number of times where both bits are 0. The sum of how two bit strings match up is then divided by N to arrive at a similarity measure with a $[0, 1]$ range.

$$s = \frac{c + d}{N}. \tag{2.8}$$

There can only be $N + 1$ possible values however. This makes for a very course grain discrete similarity measure as compared to the Tanimoto similarity ($O(N)$ versus $O(N^2)$). It can be used in the limited context where the bits turned on not in common in A and B can be ignored.

2.6.1.4   The Binary Euclidean Measure

The Euclidean distance in Equation 2.1 can be made into a convenient dissimilarity measure by the following:

$$d = \frac{\sqrt{\sum_{i=1}^{N} (A_i - B_i)^2}}{N}, \tag{2.9}$$

where A and B binary bit strings have been substituted for the vectors, $\mathbf{x}$, and $\mathbf{y}$ respectively, and i refers to the ith bit, and N is the length of the bit strings in question. This is of course equivalent to

$$d = \frac{\sqrt{\sum_{i=1}^{N} |(A_i - B_i)|}}{N} = \frac{\sqrt{\sum_{i=1}^{N} XOR(A_i, B_i)}}{N}, \tag{2.10}$$

where XOR returns 1 and 0 for true and false respectively. When implemented the latter expression may be the most efficient. Its similarity expression, known as the *Minkowski measure*, in terms of the symbols used to describe the Tanimoto measures is simply,

$$s = \sqrt{\frac{a + b}{c + d}}, \tag{2.11}$$

where $c + d$ of course equals N.

Like the simple matching coefficient, its similarity compliment, there are only $N + 1$ possible values. The Euclidean measure is a dissimilarity with a range of $[0, 1]$. It has the opposite bias of the Tanimoto, making bit strings with fewer bits on appear to be more similar than the counterpart—bit strings with many bits turned on. Researchers have noted that taking the product of the Soergel and the squared Euclidean dissimilarity [35] may help to ameliorate the biases of both. However, this product does not increase the number of possible values much beyond that of the Soergel (Tanimoto) in practice.

2.6.1.5   The Binary Cosine or Ochai Measure

The Cosine measure is a metric and, like the Euclidean measure, is also a dissimilarity. Its compliment similarity measure is known as the Ochai similarity measure. The Cosine is essentially the normalized dot product of two binary vectors.

$$d = \frac{c}{\sqrt{(c+a)(c+b)}}. \tag{2.12}$$

The cosine measure is a metric and can generate a great many more possible dissimilarity values than the Soergel, but it is nevertheless discrete. The number of possible values as a function of N is unknown. Empirically it can be shown to grow faster than the $O(N^2)$ bound of the Tanimoto measure. This fact makes this measure robust with respect to clustering problems with discrete measures, providing the size of N is relatively large, namely greater than $N \geq 2^{10}$. With this size of N, the number of possible values approaches or exceeds the number of possible values with six digit floating point precision within the range of $[0, 1]$ (namely, a million or more values).

2.6.1.6 The Hamann Measure

The Hamann similarity again is concerned with the bits turned on in common, but also the bits turned off that are in common. Let a in this instance be the count of bits in common turned off in both A and B. The Hamann similarity measure it then defined by

$$s = \frac{c}{(c+a)}. \tag{2.13}$$

Again, the Hamann measure is a metric and it can easily be seen that the number of possible values is the F_N, the Farey sequence N. The distribution of the values however differs.

2.6.1.7 Other Binary Measures

In cheminformatics, similarity searching using structural fingerprints is a common drug discovery activity. The measure of success of such searching is often based on some response property of the compound (activity, inhibition, etc.). More complicated and sophisticated binary measures have thus been explored. These measures, without loss of generality, can be used for clustering as well. They are often less efficient than the simple measures shown above, though their discrete nature is often more fine grain, so there is a tradeoff in their use for clustering [61, 71, 72, 116, 143, 150]. Here are more common association coefficients not already mentioned, ranked loosely in their computational complexity.

- Russell-Rao

$$s = \frac{c}{n} \tag{2.14}$$

- Mean Manhattan

$$s = \frac{(a+b)}{n} \tag{2.15}$$

- Dice

$$s = \sqrt{\frac{2c}{a+b+2c}}. \tag{2.16}$$

- Rand

$$s = \sqrt{\frac{c+d}{a+b+c+d}}. \tag{2.17}$$

- Forbes

$$s = \frac{cn}{(c+b)(c+a)} \tag{2.18}$$

- Folkes/Mallows

$$s = \frac{c}{\sqrt{(c+a)(c+b)}}. \tag{2.19}$$

- Fossum

$$s = \frac{n(c-1/2)^2}{(c+b)(c+a)} \tag{2.20}$$

- Simpson

$$s = \frac{c}{min((c+b),(c+a))} \tag{2.21}$$

- Kulczynski-2

$$s = \frac{\frac{c}{n}(2c+b+a)}{(c+b)(c+a)} \tag{2.22}$$

Common correlation measures, again loosely in order of their computational complexity, are:

- Yule

$$r = \frac{cd-ba}{cd+ba} \tag{2.23}$$

- Dennis

$$r = \frac{cd-ba}{\sqrt{n(c+b)(c+a)}} \tag{2.24}$$

- Pearson

$$r = \frac{cd-ba}{\sqrt{(c+b)(c+a)(b+d)(a+d)}} \tag{2.25}$$

- Stiles

$$r = \log\left[\frac{n\left(|cd-ba|-\frac{n}{2}\right)^2}{(c+b)(c+a)(b+a)(a+d)}\right] \tag{2.26}$$

In general Russell-Rao and simple matching coefficient typically perform more poorly in comparing similarity of binary strings than any of the other association or correlation measures mentioned above. Of course there may be data and application specifics that warrant the use of one or a number of the associations or correlations over the others. For instance, in the similarity searching application, the size of the data are such that the computation of some of the more complicated correlation measures can be tolerated, they may perform marginally better, with the same data and application caveat. Since both similarity searching and clustering have much the same premise of structure-biological activity, the binary measures are often transferable. Some of the measures are also used in cluster validity studies, such as the Rand, Folkes/Mallows, Tanimoto (more commonly called the Jaccard in cluster validation), and the Euclidean measures.

2.6.2 Count Data Measures

Count measures can be derived from both binary and continuous measures. Thus, for example there is an analogous Tanimoto measure for count data (count fingerprints); and the Euclidean distance can be used with the count data type, by converting the count data into real valued numbers (e.g., assign 4.0 to the count value 4). The discrete nature of the values does not relate to the range of possible counts. In the discussion above concerning the counts of subgraph matches of a single fingerprint subgraph key, it was noted that the typical subgraph key would rarely have more than just a few matches within a single compound. Thus, say for a database of several million compounds, if no key has more than some small number of matches (e.g., $<< 100$), the number of possible similarity values will not increase by a great deal over the binary case for the measure in question. Indeed, it was noticed that for this and other reasons, the results one obtains for count data concerning both clustering and similarity searching is often only marginally different from those obtained with the equivalent binary data.

2.6.2.1 The Tanimoto Count Measure

The count Tanimoto similarity is defined as follows, where N is the dimension:

$$s = \sum_{i=1}^{N} \frac{a_i}{b_i}, \tag{2.27}$$

where a_i is the lesser of the ith count from either of what are now defined as count vectors, A and B, and b_i is the greater. If b_i is 0, then that term of the sum is 0. This similarity cannot be transformed into a dissimilarity metric that satisfies the triangle inequality. The values of this expression are also not necessarily within a range of $[0, 1]$. Normalizing the values to $[0, 1]$ is also not necessarily straight forward, as it can skew the distribution greatly. This can be seen by a normalization that takes the maximum count value for all vectors, K, as the denominator, which seems the most natural normalization.

2.6.2.2   The Cosine Count Measure

Here the cosine measure is the actual angle as defined by the arccos of the normalized dot product of the two count vectors. This is a discrete form of the continuous measure as it would be defined to generate the angle between the two vectors if they were indeed continuous.

2.6.3   Continuous Data Measures

The most natural continuous data measure is the Euclidean distance, or the L_2 norm. The terminology of L_2 *norm* comes from the general class of metrics known as the Minkowski metric, defined as:

$$L_n(\mathbf{x}, \mathbf{y}) = \sum_{i=1}^{N} (|\mathbf{x_i} - \mathbf{y_i}|^{\mathbf{n}})^{(1/\mathbf{n})}, \qquad (2.28)$$

where N is the dimension, and $n \geq 1$. Aside from any issues of precision, problems arising from discrete types are not an issue. Other norms can be used such as L_1 (the *City Block* or *Manhattan* distance) and L_∞, though most common uses of the Minkowski metric are of the Euclidean form. There may however be special cases where specific data applications might warrant the use of other Minkowski metrics. Modifications of the Euclidean distance concern how variables are possibly weighted or normalized. Weighting for issues concerning noise and variable correlation also drive the use of other methods for determining a measure among continuous values, such as the use of the Mahalanobis measure, the use of principal components analysis, or the use of the Pearson correlation coefficient to derive a useful dissimilarity measure between data elements for clustering. These three methods and especially latter two are more in line with transformations of the data.

2.6.3.1   Continuous and Weighted Forms of Euclidean Distance

$$d = \sqrt{\sum_{i=1}^{N} (x_i - y_i)^2}, \qquad (2.29)$$

where $\mathbf{x}$, and $\mathbf{y}$ are now continuous valued vectors, and i refers to the ith element in the respective vector. The Euclidean distance can be weighted if there is expert knowledge concerning the relative contribution of each variable, as expressed by a vector of weights $\mathbf{a}$.

$$d = \sqrt{\sum_{i=1}^{N} a_i (x_i - y_i)^2}. \qquad (2.30)$$

If the absolute variation of each variable differs greatly, the Euclidean distance can be normalized to this variation by the following:

$$d = \sqrt{\sum_{i=1}^{N} \frac{(A_i - B_i)^2}{\sigma_i^2}}, \qquad (2.31)$$

where σ_i^2 is the variance of the ith variable.

2.6.3.2 Manhattan Distance

The Manhattan Distance, sometimes called the city block, or taxicab metric, defines the shortest distance given a grid, hence the name. More formally, it is defined by

$$d = \sum_{i=1}^{N} |x_i - y_i|. \qquad (2.32)$$

This measure is sometimes used for faster computation on large data sets or for special feature variables.

2.6.3.3 L_∞ or Supremum Norm

L_∞ or Supremum Norm is a curious metric in that it simply defines the distance as the maximum difference between the respective features of the vectors x and y: it is defined by

$$d = max\left(|x_i - y_i|\right), \qquad (2.33)$$

for $i = 1, 2, \ldots, N$, where N is the dimension of the vectors. Metrics like the Manhattan and the Supremum Norm are fast to calculate, so if the data and application warrants, these metrics can at times be useful, when calculating more complicated metrics or measures is out of the question due to very large data sizes.

2.6.3.4 Cosine

The cosine similarity is a common measure that can be defined by its usual vector dot product form,

$$s = \frac{\mathbf{x} \cdot \mathbf{y}}{||\mathbf{x}||\,||\mathbf{y}||} \qquad (2.34)$$

It defines the cosine of the angle between the two vectors $\mathbf{x}$ and $\mathbf{y}$. This measure can be very useful in a number of different settings, as can, by extension, simply the dot product.

2.6.3.5   Pearson Correlation Coefficient

Correlation coefficients are often used in clustering and the most ubiquitous is the Pearson sample correlation coefficient:

$$r = \frac{\sum_{i=1}^{N}(x_i - \hat{\mathbf{x}})(y_i - \hat{\mathbf{y}})}{\sqrt{\sum_{i=1}^{N}(x_i - \hat{\mathbf{y}})^2}\sqrt{\sum_{i=1}^{N}(y_i - \hat{\mathbf{y}})^2}}, \tag{2.35}$$

Its range is from $[-1, 1]$, namely, from negative correlation to positive correlation, and it is invariant to linear transformations and scaling. The question becomes whether if two data objects are negatively correlated they are somehow further or closer than if they had no correlation. The use of the absolute value of the correlation assumes that negative and positive correlation are the same. This is certainly plausible, given certain data. Say, for example, the data are a set of times series. One could then ask are time series that do the opposite from one another at every time step (think of a sawtooth function in opposite phase) closely related, or at the limits of correlation. This would depend on the application, of course.

2.6.3.6   Mahalanobis Distance

The Mahalanobis is also invariant to scale but really is a distance between a single data point to a set of data points for which the mean and covariance matrix are known. The principal components of the point set represent the extent of a hyper-ellipsoid in N dimensional space. And it is that extent of the point set that the distance between the single data point and the point set is determined. While the distribution of the point set may not be ellipsoidal at all, the distance calculation assumes that it is. Where data sets have gaussian or gaussian-like distributions, this distance is more accurate. Assume the mean of the point set, S, and its covariance matrix are known, $\bar{\mu}$, and Σ, respectively, then the Mahalanobis distance to the point $\mathbf{x}$ is:

$$r = \left((\mathbf{x} - \bar{\mu})^{\mathbf{T}}\Sigma^{-1}(\mathbf{x} - \bar{\mu})\right)^{1/2}, \tag{2.36}$$

where r is the effective radius about the ellipsoid (or hyper ellipsoid if the dimension is greater than 3) defined by the eigenvectors and eigenvalues of Σ, and upon which $\mathbf{x}$ resides. Any $\mathbf{x}$ on this contour about the ellipsoid has the Mahalanobis distance r.

If the points lie in the same distribution and we want to create a dissimilarity measure between two data points, we can again use the covariance matrix, Σ, of the entire set and arrive at the following:

$$d = \left((\mathbf{x} - \mathbf{y})^{\mathbf{T}}\Sigma^{-1}(\mathbf{x} - \mathbf{y})\right)^{1/2}, \tag{2.37}$$

However, the data points need not be a cloud of points whose distribution necessarily is gaussian, but the more the distribution deviates from gaussian the value of the distance can become more problematic. Nevertheless, such dissimilarities are sometimes used to cluster data. One feature of the Mahalanobis distance is that it is invariant to translation and scale. This may serve in some clustering settings.

2.6.4 Mixed Type Data

There are very few algorithms that can operate on mixed data as is. In most instances, clustering data of mixed type requires transformations to a single data type that takes into account the issues of the various types in question, the goal of the cluster analysis, and expert knowledge of the data. Divisive monothetic clustering is a recursive partitioning method that can been shown to operate on mixed data types to cluster the data variables as oppose to clustering the data items [26, 24].

2.7 Proximity Matrices

Proximity matrices have properties that may impact on the form of analysis. For example, a proximity matrix may be symmetric or asymmetric, depending on the measure used to generate it. If symmetric, depending on the measure used to generate it and the data, it may or may not have all positive eigenvalues. In each of these examples, the properties of the matrix will likely determine the type of transformations to visualize the groups (such as Principal Components Analysis (PCA), Isometric or Classical Multi-Dimensional Scaling (MDS)) or clustering algorithms that might be applied to this data. In addition, the properties of proximity matrices that are composed from several proximity matrices will determine what operations can and cannot be applied to these matrices for either clustering or visualization.

2.8 Symmetric Matrices

Equation 2.38 defines a symmetric matrix $\mathbf{S}$:

$$\mathbf{S} = \frac{(A_{i,j} + A_{j,i})}{2},\tag{2.38}$$

where i and j are the pair of items being compared, and $A_{i,j}$ is the similarity measure between item i and item j. $A_{j,i}$ is the measure in the reverse comparison. For a symmetric matrix

$$(A_{i,j}) = (A_{j,i}) \tag{2.39}$$

In Figure 2.9 a symmetric proximity matrix is displayed for a set of Cox-2 ligands, encoded as binary fingerprints with 768 bits and compared with the dissimilarity compliment of the Tanimoto measure. The inhibition of the Cox-2 enzyme decreases pain and inflammation and is a well studied set of compounds. The proximity matrix displays the pairwise similarities across the data set utilizing the Tanimoto similarity measure. Each square represents one pairwise comparison. The intensity of the greyscale reflects the Tanimoto scale of 0 to 1. A value of 1 corresponds to black and a 0 value corresponds to white. Along the diagonal as a structure's fingerprint is compared to itself, one would expect to see black and as one moves off the diagonal, the grey scale reflects the similarity to other compounds in the data set. The data are partially

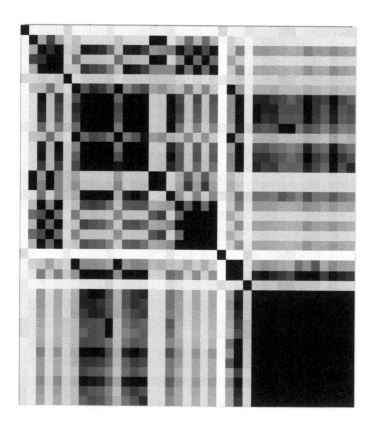

FIGURE 2.9: Example symmetric proximity matrix for a set of structures which display activity for Cox-2 (an enzyme responsible for pain and inflammation).

ordered by chemical series (the initial SMILES encodings of the compounds were lexigraphically sorted), and one does observe pockets of black squares where similar compounds happen to be displayed next to each other.

2.8.1 Asymmetric Matrices

An asymmetric matrix, $\mathbf{A}$, derived from asymmetric measures can be decomposed into a symmetric matrix $\mathbf{S}$ and a *skew symmetric matrix*, $\mathbf{N}$:

$$\mathbf{A} = \mathbf{S} + \mathbf{N}, \tag{2.40}$$

where the symmetric matrix is defined as,

$$\mathbf{S} = \frac{(A_{i,j} + A_{j,i})}{2}, \tag{2.41}$$

and the skew symmetric matrix is defined as,

$$\mathbf{N} = \frac{(A_{i,j} - A_{j,i})}{2}. \tag{2.42}$$

An asymmetric measure may or may not produce necessarily much asymmetry. Going back to the social science example of finding individual subgroups of like-minded people, it may turn out that everyone or almost everyone has equivalent pair-wise preferences for the other individuals in the group, in which case there may be very little or no asymmetry in the pairwise preferences. One could easily envision the opposite case as well, where there is a lot of pairwise asymmetric preference. In the use of asymmetric matrices for clustering, since there is only a limited number of asymmetric clustering algorithms, knowing the level of the asymmetry would be advantagious. If the level of asymmetry is negligible, the asymmetric matrix can be transformed into a symmetric matrix using Equation 2.16 above. There is a measure of the percent asymmetry, P, that can be used to show this:

$$P = \frac{\sum_{i,j} \mathbf{N} \otimes \mathbf{N}}{\sum_{i,j} \mathbf{N} \otimes \mathbf{N} + \sum_{i,j} \mathbf{S} \otimes \mathbf{S}} \times 100, \tag{2.43}$$

where $\otimes$ is the Hadamard product, the element-wise product of two matrices, and the sum is over all elements of the resulting product matrix.

In addition to finding the percent asymmetry, the asymmetric drift can be calculated and displayed via a MDS plot. The component of drift is obtained by calculating the MDS coordinates of the symmetric matrix $\mathbf{S}$, where $a_{i,j}$ is the difference between the ith and jth MDS component vectors. The component of drift then is

$$N_{i,j} \frac{a_{i,j}}{\|a_{i,j}\|}, \tag{2.44}$$

where $N_{i,j}$ is the i,jth element of the skew matrix $\mathbf{S}$. Figure 2.10 is an MDS plot displaying asymmetric drift for a set of ligands. The asymmetric drift is represented by the arrows. The green dots represent the MDS placement of the symmetric proximity matrix and the red circles represent a scaled variable property.

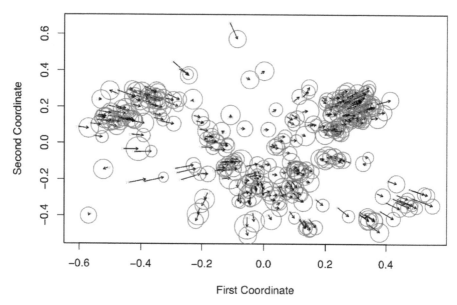

FIGURE 2.10: Example asymmetric drift plot. The centers of the circles represent the symmetric placement in MDS, the arrows represent the asymmetric drift, the circles represent a scaled property.

2.8.2   Hadamard Product of Two Matrices

Hadamard product matrices have other uses in clustering aside from exploring asymmetric matrix properties. Symmetric matrices can be used to compose other symmetric matrices. For example, matrices A and B may be matrices generated different dissimilarity measures over the same data. In one such example, the molecular shape Tanimoto and the electrostatic Tanimoto matrices can be composed with the Hadamard product, and once suitably normalized, the result can be used as a proximity matrix for clustering. The composition of these two measures can be made via the Hadamard product, but properties of both matrices and the Hadamard product thereof may determine what possible further clustering algorithms or statistical analysis such as PCA or MDS can be performed. If, say, the two matrices are positive semi-definite, the Hadamard product of the two matrices is not necessarily so.

2.8.3 Ultrametricity

Another important matrix property, especially for clustering, is ultra-metricity. An ultrametric matrix is defined such that of the three pairs of i, j, and k, two such pairs are equivalent in value and are at the maximum of the three pairs. This property can be used to evaluate the tree structures of hierarchical algorithms and is directly related to clustering monotonicity, discussed in Chapter 6 with regards to hierarchical structures. The matrix of merged values in the generation of hierarchical clustering structures that maintain the ultrametric properties are called the *cophenetic* matrix. The cophenetic matrix can be compared to the original proximity matrix to form a correlation measure via the Mantel statistic, the element-wise correlation of the matrices. This is known as the cophenetic correlation, where values near one suggest that the hierarchical structure is consistent with the groups so formed given the proximity matrix.

2.8.4 Positive Semidefinite Matrices

In a number of settings such as MDS and PCA, whether a matrix has the property of being positive semidefinite has important consequences. For real valued matrices, a matrix A is said to be positive semidefinite if

$$\mathbf{x}^T A \mathbf{x} \geq 0. \tag{2.45}$$

2.9 Dimensionality, Components, Discriminants

The feature space of a data set may be complex and large. It may be that certain features are very important to the grouping of data items and others are of little or no importance, and, indeed, may be largely noise or otherwise confounding. Efficient linear analytical methods exist for reducing the set of features to a smaller set of features, or a smaller linear combination of the features. These can be used to explore the data as a set of projections, or even to use the small or transformed set of features for clustering. The latter has the caveat that such linear selection or transformations may not be optimal for the natural grouping of the data.

The most common method of feature selection and transformation is PCA. It is widely used in exploratory analysis, regression, classification, and, occasionally, cluster analysis. Assume in the simple case that the data set features are continuous valued and we can treat them in a Euclidean space. We can then find a linear combination of features such that they form an axis through the set of data points where the points have the greatest variance. That is, if we project the data points along this axis, the points are in a configuration of their largest variance. This is the linear first, or principal component of the data. The second principal component is that component that represents the next greatest variance in the data, that is orthogonal to the first principal component. The remaining orthogonal principal components are determined similarly. There are two ways of generating principal components: directly on the data matrix using singular value decomposition, and via the covariance matrix of the data. Note, that the second principal component is orthogonal to the first, and not necessarily the second greatest variance in the data aside from the principal component. There could indeed be a basis of axes that are not orthogonal representing the sequential largest variances. There are methods that relax the orthogonality constraint to find such a basis. Multidimensional scaling attempts to project the similarities or dissimilarities between the data items into a smaller set of dimensions than that of the total number data features. These methods are typically used to visualize the data in two or three dimensions and try to approximately preserve the similarities or dissimilarities from the larger dimension. There are numerous such functions that make these projections, each with properties that tend to accentuate certain aspects of the data. Some MDS methods are linear, such as classical MDS, that uses an eigenvalue decomposition of the dissimilarity matrix, very much akin to PCA.

In a sense, clustering criterion functions strive for discrimination where the classes are unknown. Discrimination methods are not necessarily in the domain of cluster analysis, but they are important to concept to understand from a post facto analysis standpoint. If we knew of the classes of the data before hand, such as in classification modeling, we might want to find the optimal linear separation of the classes in the feature space. Since this is unsupervised clustering this is not the case, but equally, say we have clustered a data set and we want to perform this same operation on the groups found with a clustering method. We can use Fisher's linear discriminant method to do just that. We may wish, for example, to visualize the groups found via clustering with discriminant surfaces, whether linear or nonlinear, delineating the groups. These methods are most effective with just a few classes however, so their utility is somewhat limited for cluster analysis. But more importantly it is good to have a sense of discriminant functions since they are so closely related to cluster criterion functions in delineating groups.

2.9.1   Principal Component Analysis (PCA)

PCA is the linear interpretation of the greatest variance in feature space of the data in orthogonal terms. The variance in feature space is contained in the covariance matrix for the data under consideration.

2.9.1.1   Covariance Matrix

The Covariance Matrix is defined in Equation 2.46.

$$\Sigma = \frac{(\mathbf{x} - \bar{\mu}) \bullet (\mathbf{x} - \bar{\mu})^{\mathbf{T}}}{n} \qquad (2.46)$$

The numerator is the outer product of the data minus its means, divided by n, the number of data elements.

$$A^1 \Sigma A = D \qquad (2.47)$$

D is the diagonal matrix of the eigenvalues. The columns of A are their corresponding eigenvectors. The corresponding eigenvectors of the rank ordered eigenvalues are the principal components.

2.9.1.2 Singular Value Decomposition

Principal components can also be calculated using singular value decomposition.

$$\mathbf{X} = \mathbf{UDV^T} \qquad (2.48)$$

X is the data matrix, U is the left singular vectors, D is a diagonal singular value matrix, V contains the right singular the orthonormal eigenvectors. The PCA using SVD is utilized in the analysis of gene expression data [142].

2.9.2 Non-Negative Matrix Factorization

$$\mathbf{X} = \mathbf{WH} + \mathbf{R} \qquad (2.49)$$

$\mathbf{WH} + \mathbf{R}$ is the *Non-negative Factorization* of the data matrix X. It is related to PCA in that it forms a basis of vectors, and also very much like K-means clustering where W contains the cluster centroids and H contains the cluster membership vector. W and H are non-negative, but R is a residual matrix which may contain negative values [34, 156].

2.9.3 Multidimensional Scaling

There are numerous forms of multidimensional scaling [16, 31], often used for visualization, but also on occasion, dimension reduction. *Classical Multidimensional Scaling* is using PCA on the data as above, where we want to use just a small number of the first principal components to project the data relationships into a small dimensional space. Equation 2.48 or Eq. 2.46 covariance eigenvalue/eigenvector methods can be utilized. Alternatively, a dissimilarity or distance matrix can be used in a non-linear fashion with numerous techniques to arrive at a projection into a small number of dimensions. Many of these forms of MDS are based on some form of force directed methods, much like spring embeddings for graph layouts, and are iterative in nature and typically find locally optimal solutions. These methods can be somewhat slow on large data sets, but for modest sized data sets they can perform quite well in revealing structure in the data.

2.9.4   Discriminants

Linear, quadratic, and other non-linear discriminants are techniques used in supervised learning, but such techniques are conceptually useful in understanding clustering methods and validation. Discriminants create a set of decision surfaces that distinguish classes. In the ideal, a clustering would produce groups that would conform to some form of decision surface separating the groups with the least misclassification. The linear optimal discriminant is known as Fisher's discriminant.

2.9.4.1   Fisher's Linear Discriminant

Fischer's linear discriminant is a classification method in which multi-dimensional data are projected into typically two classes (though it can be generalized) by minimizing the variance within each class and maximizing the distance between the means of the two classes. The goal of Fischer's is to create a linear decision surface between classes. Equation 2.50 is one form of the Fisher's linear discriminant.

$$\max \mathbf{F(p)} = \frac{|\bar{\mu}_1 - \bar{\mu}_2|^2}{\sigma_1^2 + \sigma_2^2}, \tag{2.50}$$

where σ_1^2 and σ_2^2 are the variances in each of the two classes for a given decision surface p, $\bar{\mu}_1$ and $\bar{\mu}_2$ are the means of each class. This linear surface forms the optimal separation between two classes with the least misclassification. (For an analogous linear separation, see Figure 4.1 in Chapter 4.)

2.10   Graph Theory

Graphs are combinatorial objects with more or less unlimited algorithms and practical applications. They are used to model many applications within bioinformatics and drug discovery, and are common throughout the study of cluster analysis. A very brief introduction is offered here, but readers would not be remiss in searching out and studying graph theory and graph algorithms [15, 151].

A simple graph has vertices (or nodes) and edges, very much like a transportation network of cities and roads respectively. Weights or labels can be applied to nodes and edges: in the case of cities and roads, population and distance are example weightings of nodes and edges respectively. A chemical graph is slightly more complicated form of a simple graph—simply replace cities and roads with atoms and bonds. Atoms can have numerous labels and properties—a nitrogen atom has the label N and the property of being charged or not, etc. There are single, double, and triple labeled bonds. A portion of a graph—a proper subset of vertices and their associated edges—is known as a subgraph. In a chemical graph there may be important subgraphs (often referred to as substructures) that have labels such as a benzene ring or a phosphate group. Figure 2.11 shows two views of the simple undirected graph of the caffeine molecule depiction in Figure 2.1. The graph depiction in Figure 2.11a shows the caffeine molecule without the atom labels but retaining the chemical bond information and the ring symmetry layout. Figure 2.11b shows the underlying graph without the chemical bond information aside from single edges. It is also laid out on the page quite differently, with little concern for ring symmetry. Fundamental graph properties of both graphs are the same as are the results of algorithm procedures performed on both graphs, label information notwithstanding.

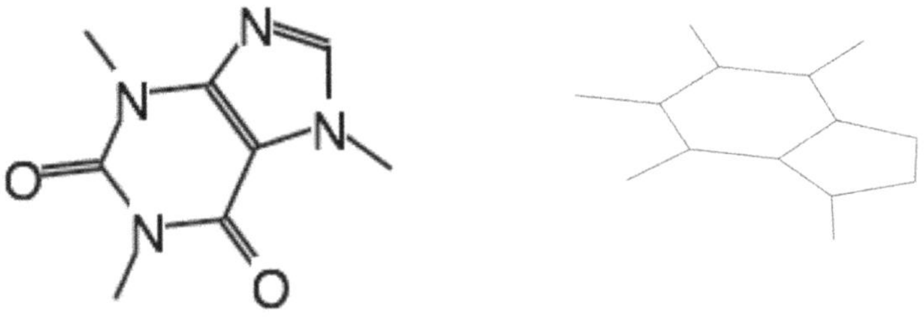

(a)   (b)

FIGURE 2.11: Simple graph of a caffeine molecule: a) with just bond information; b) simple graph without any chemical information.

It turns out that nearly all small molecules can be represented as *planar* graphs. That is, they can be laid out on a two dimensional surface without edge crossings. The layouts are also called *graph embeddings*. It can also be shown that planar graphs can even have straight line embeddings in two dimensions [33, 44, 117]. Both the figures of the caffeine graph are displayed with planar embeddings on the page. Note that planar means any two dimensional surface, so that the planar embeddings can be on spheres, toruses, etc. Planarity of graphs often is important in the computational complexity of graph algorithms, such that if it is known that a graph is planar there may be an algorithm that operates on planar graphs that is more efficient than a non-planar graph, namely, the difference between an algorithm in polynomial time versus in one in *NP*.

Graphs play an important role in cheminformatics from the simple standpoint of chemical graph theory, but graphs and graph algorithms also are important in cluster analysis algorithms as numerous clustering algorithms are fundamentally graph algorithms or portions of the algorithms can be modeled as graph problems. For example, the proximity matrix can be considered, and often is, a weighted graph. Also, the results of hierarchical algorithms are displayed as a graph: namely, as a simple connected graph without cycles. Such graphs are called *trees*. There are general and binary rooted trees, and unrooted or free trees. A hierarchy is a rooted tree, and in many instances these trees are binary, but they need not be. Simple tree concepts are important in clustering algorithms that generate hierarchies and the analysis and exploration of those hierarchies. Dendrogram depictions of clustering hierarchies are almost always binary trees. Figure 2.12 shows several typical tree graph representations and layouts. Rooted trees are laid out in a hierarchy and look more like a root system of a tree that a standing tree per se. Occasionally they can be laid on their side as in Figure 2.12b. Trees need not be rooted, nor binary and their layouts on a page can have many representations. Since hierarchies can be represented by a tree, how one operates on a tree (searching, decomposition, re-arranging, etc.) can be translated directly to how one operates on a hierarchy. A set of rooted trees is called a *forest*. Thus, for example, given a rooted tree, a set of distinct *subtrees* in the tree represents a forest of trees.

Graph properties represent some overall feature of a graph. For example, a tree is a type of graph with a specific structure: e.g., there are no cycles in a tree graph. Graphs have general properties such as the graph's radius, diameter, circumference, eccentricity, biconnectivity, k-connectivity, etc. These are often referred to in cheminformatics as topological properties, and they are legion - new ones can be defined at will. Properties of a graph may also play a role in the computational complexity of a graph algorithm operating on that graph. Worst case performance may depend on very special properties of a graph, that may be quite rare, such that the average case performance - and what is often the case in practice - can differ substantially.

Simple graphs are also called binary graphs or undirected graphs. Figure 2.13 displays an undirected labeled weighted graph. Edge weights may represent proximities or capacities, etc. There are also binary directed graphs, where each edge is an *arc* with a specific direction, from one node to another. Directed graphs will figure prominently in the discussion of asymmetric measures. A multigraph is a graph with more than one edge between nodes and they may have self-loops - an edge that returns to the same vertex. Chemical graphs can be considered multigraphs with bonds as multiple edges, but without self-loops. However, typically chemical graphs are thought of as simple graphs where single, double, and triple bonds are considered edge labels.

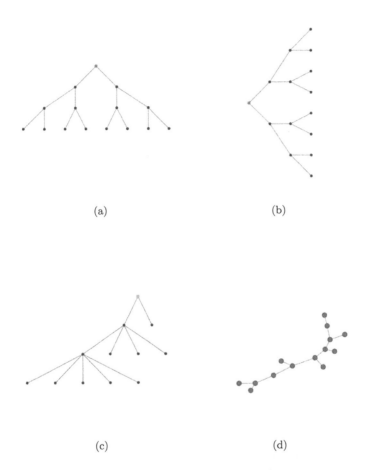

(a)

(b)

(c)

(d)

FIGURE 2.12: Simple rooted tree graph and unrooted or free tree: a) rooted binary tree; b) tree in *a* rotated; c) general rooted tree; and d) unrooted or free tree.

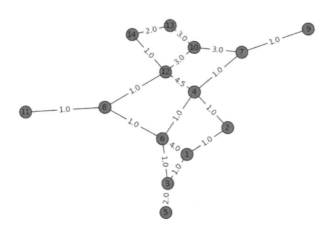

FIGURE 2.13: This is a figure of an undirected labeled weighted graph.

A hypergraph contains edges that can contain more than two nodes. Multi-graphs and hypergraphs can also be either undirected or directed. In the case of hypergraphs, the direction is in the form of a sequence of nodes, versus the set of nodes in an undirected hypergraph. The more complicated graphs such as multigraphs and hypergraphs are more rare in the discussion of cluster analysis, though hypergraphs are a way to visualize overlapping clusters. For example, a Venn diagram is one way to visualize a hypergraph or, equivalently, a collection of overlapping clusters. Though if there are a great many clusters and a good deal of overlapping, this often becomes visually unmanageable. For simple hypergraphs however the Venn diagram and other methods can be used to visualize overlapping clusters as discussed in Chapter 10.

An important aspect of designing efficient graph algorithms is how to store the graph or graphs. There are several common methods for storing a graph such that it can be operated on efficiently. The most common is the adjacency list (the connection table of a molecule is a form of an adjacency list), which is a list of nodes, each with a list of nodes with which it has an edge. This is the most efficient storage, and often the most efficient to use in an algorithm. Some algorithms however need what is known as an adjacency matrix. An adjacency matrix is a matrix where each edge is represented by an (i,j) pair of nodes in the matrix. If there is no edge between node i and node j, a 0 is placed in the matrix, otherwise a 1 or a weight or label. There are other less common data structures such as an ordered pair, or edge adjacency list. A proximity matrix, whether symmetric or asymmetric can be thought of as a adjacency matrix for a simple binary graph or simple directed graph respectively. Various graph algorithms can then be used to operate on these matrices either for clustering purposes directly or for visualization of clustering results.

Graph algorithms can be very powerful tools, and often if a problem can be transformed to a graph problem, it may lead to the design of a more efficient algorithm for that problem. An example can be found in network flow algorithms that were originally solved via linear programming methods of the 1950s and 1960s whose computational complexity was in the worst case extreme. Transformed to a graph problem, and with the use of graph algorithms, far more efficient algorithms were designed both in theory and in practice in the 1980s and 1990s. This is an important notion to keep in reserve: if there is a clustering method that is an iterative, gradient search-like algorithm, that approaches an optimal solution, there may be a more efficient graph algorithm, more combinatorial in nature, that in practice outperforms such an approach. Physicists, engineers, and chemists often approach problems from the standpoint of analogue processes, and model them with analogue-like algorithms. Machine learning algorithms for artificial intelligence—often based on neural processes—also have this flavor. Some clustering algorithms in this book that derive their provenance from those fields will thus naturally have that same flavor, and they may work well and appropriately for the data at hand, but taking a more graph theoretic, combinatorial approach may work just as well or better, and more efficiently. After all, the computers used to generate clustering results are not analogue machines. A recognition of this fact is often important in the understanding of clustering ambiguity and practical efficiency of clustering algorithms.

2.11 Glossary

adjacency list: A list of lists of edge-vertex relationships.

adjacency matrix: Full matrix of edge-vertex relationships.

arc: A directed edge.

centroid: The geometric center of a group.

centrotype: The center or representative of a group, that may not be the centroid strictly speaking.

directed hypergraph: Edges are an ordered sequence of vertices(nodes).

directed multigraph: A multigraph with arcs.

edge: Equivalence relation between two vertices

exemplar: Centrotype

hypergraph: Edges are a set of vertices(nodes).

metric measure: Adheres to the triangle inequality.

multigraph: May include multiple edges between nodes, or a node may have an edge to itself.

node: A vertex.

proximity matrix: Matrix of dissimilarity (may be a distance) or similarity relationships.

simple binary graph: Vertices and undirected edges.

simple directed graph: Vertices and directed edges (arcs).

subgraph: A portion(a proper subset) of a graph.

2.12   Exercises

1. Create a set of distinct random bit strings of length 10. Calculate all pairs similarity measures between the strings with normalized euclidean, tanimoto and 1—cosine. Plot the similarity distributions. Compare the results.

2. Create a set of distinct count fingerprints of length 10, where the counts range from 0 to 8. Calculate all pairs similarity measures between the strings with the count tanimoto, 1—cosine, and the normalized euclidean similarity measures. Compare the distributions of the similarity values to those distributions calculated in Exercise 2.1.

3. Prove that the count tanimoto is not a metric, either by an existence proof or analytically.

4. Take a very small dataset greater than 5 members and less than 10 members, generate a symmetric similarity matrix, plot the distribution. Pick three thresholds from the distribution. Draw the weighted graph up to each threshold. Is there a threshold where the graph remains planar?

5. Take any data set that can be compared with a metric similarity and generate the principal components and plot the ranked eigen values. Plot the data points in the first two principal components. Plot several combinations of the first five principal components and discuss the results [133].

Chapter 3

Clustering Forms

There are two basic forms of clustering algorithms: partitional and hierarchical. In their simplest form, partitional algorithms take as input the data set with N members, and a number, $k << N$ and, in all likelihood, $k > 2$, and return a partition of the data set into k disjoint subsets that represent the clusters of the data set. There are however a vast number of possible partitions for large N and non-trivial k. Even if $N = 50$ and $k = 5$, there are 50 *choose* 5 possible partitions: there are over 2 million such partitions. To get a sense of the explosive nature of the number of possible partitions, say, we have 60 items and think that there are 15 clusters: there are over 53 trillion possible distinct partitions to search exhaustively to find the optimal one.

Almost surely we will not know exactly what number of clusters there should be and we will want to test across some range of k. We can look at the total number of possible partitions over all of k of a set with N members to get a wild upper bound on the number of partitions we may have to explore. This is given by the Bell number [141, 55, 115], i.e., the sum of the Sterling numbers of the second kind, N subset k, over all k, where $k = 0, 1, \ldots, N$.

$$B_N = \sum_{k=0}^{N} S(N, k) \tag{3.1}$$

The Bell number is exponential in N, so that the number of possible partitions of a data set grows very rapidly as a function of its size in N. When $N = 10$ there are already 115,975 possible partitions for all k. For most applications k is likely to be in a small range, but this still represents a exponential number of partitions with respect to N. It is important to recognize that, though partitional algorithms are not designed to search all possible $S(N, k)$ for the optimal partition under some criterion, the partition space is very large and that such algorithms may only return a suboptimal solution for non-trivial k.

Hierarchical algorithms construct a tree, either from the top down, splitting the data set into smaller and smaller groups—known as *divisive* clustering, or from the bottom up, merging single data elements into groups and then merging larger and larger groups—known as *agglomerative* clustering. In most instances these trees are rooted binary trees, labeled at the leaves with the N data elements to be clustered. As such the tree represents, loosely, a topology of relationships between the data elements and increasingly larger groups of those elements.

In Figure 3.1, 30 data elements, enumerated by the leaves of the dendrogram, have been clustered with an agglomerative clustering algorithm—bottom up—joining first pairs of data elements, then singletons or other pairs, then larger and larger groups. The *Height* along the *y*-axis refers to the measure whereby the leaves or groups are joined (or, in the case of divisive algorithms, divided). The measure may indeed be the measure of pairwise dissimilarity among the data elements as is the case in the figure, or another value representing a quantitative merge criterion that is typically based on the original measure. The dendrogram in Figure 3.1 was created by the results of an agglomerative hierarchical algorithm, often referred to as *complete link* or *complete linkage*, but it could easily be produced by another agglom-

Generic Dendrogram or Cluster Topology

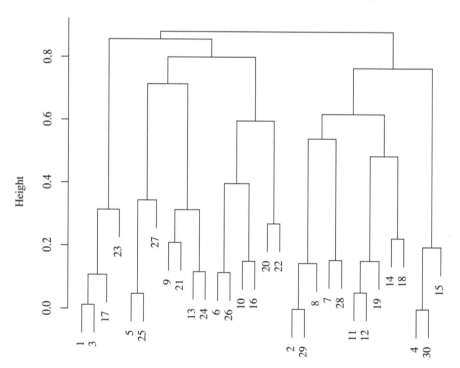

Leaves

FIGURE 3.1: A generic dendrogram of clustered data with 30 members denoted by the leaf numbering, where the dissimilarity ranges from $[0, 1]$, and the *Height* is directly related to the measure used to join the leaves and groups.

erative or divisive algorithm and look quite similar. The data set used is a set of compound fingerprints from a small group of organic compounds. A matrix of $1 - Tanimoto$ measure values between these fingerprints was used for input into the clustering algorithm. The y-axis is in units of $1 - Tanimoto$ values.

The positioning on the x-axis is one of many possible permutations, such that the tree edges don't cross. It is important to realize that the internal nodes in the tree can flip and change the permutation, so that for example, though the merged pairs $(13, 26)$ and $(6, 26)$ are close together in the dendrogram, the internal node above them at the height of 0.6 can be flipped without changing the clustering - but changing the permutation of the leaves with respect to the x-axis - and position $(13, 26)$ close to the pair $(20, 22)$, and $(6, 26)$ next to the pair $(2, 29)$, though now at different levels

Each level of the dendrogram or tree can be considered a cut in the hierarchy. There are $N - 1$ possible level cuts and therefore possible partitions in a single tree. There are other possible cuts, whereby sub-portions of the tree are cut at different levels, but this assumes that the criterion by which the groups are merged are to be in some sense ignored. Some clusters would span far different regions of the space than others given such cuts. Naturally, given the Bell number, the $N - 1$ levels, or partitions, of such dendrograms represent a very small fraction of the total number of set partitions.

Thus, whether trying to find a single partition or a hierarchy from which to choose a partition, an algorithm needs to restrict the number of possible partitions from which to draw the most meaningful partition possible in polynomial time in N. Other algorithms aside from direct partitioning and hierarchical methods are in a sense heuristics that try to derive a partition in some iterative fashion.

There are other algorithms that are some variation on one of the two basic forms, or a mixture of both. There are also classes of algorithms where cluster membership is based on a probability such as fuzzy clustering algorithms, or where data elements can reside in more than one cluster found in overlapping clustering algorithms. Self-organizing algorithms are another group of algorithms that are either partitional or hierarchical, but whose criteria for forming partitions or hierarchies are derived from self-organizing precepts.

3.1 Partitional

Partitional clustering algorithms are in some sense at the core of clustering techniques—they attempt to derive a meaningful partition of the data and arrive at a set of k clusters directly. Luckily, there are many criteria that form the basis of algorithms such that the search for the best partition can arrive at the very least an approximation to a optimal partition. The K-means algorithm is a clustering algorithm that partitions the data into clusters via an iterative procedure, that in general converges upon a solution in polynomial time. The solution may or may not be optimal; in fact, it is most likely a locally optimal solution.

The most common and popular partitional clustering algorithm is an iterative descent method known as K-means. It is easiest to understand in the context of continuous valued data in a Euclidean space. Groups are found by their nearest neighbor to some center. A group's center is then moved to its mean and the process is repeated. In a Euclidean space, when the mean is used as the measure of central tendency, this amounts to putting the groups within a *Voronoi* set of regions defined by the means. The means move and the groups change until either the solution converges such that the groups stop changing or the number of iterations is exhausted.

A simple version is shown in *Algorithm 3.1*. K groups (G) are desired and K arbitrary means (M) are selected. The means can be (pseudo or quasi) randomly selected data points, or random data in the space of the data. Nearest neighbors are found for the K centers, M, and stored in a list of groups (clusters), G. The means are then recalculated for each group, generating a new set of means. When the stopping conditions are met, the groups are output along with the respective means. Other measures of central tendency can be substituted, such as the median-like, K-medoid, or mode, K-modes, given data considerations. The algorithm is quite flexible and efficient, and there are many variations and subroutines that can improve the efficacy and performance, depending on the data and the goals of the clustering.

Algorithm 3.1 $Kmeans(Data, K, Iterations)$

1: $G \leftarrow KEmptyLists()$
2: $M \leftarrow ArbCenters(Data, K)$ //Find K arbitrary centers within Data
3: $G \leftarrow FindNN(Data, M)$ //Find the Nearest Neighbors to M
4: $M \leftarrow FindCenters(Data, M)$ //Find the new centers of the groups in G
5: $i \leftarrow 1$
6: **repeat**
7: $\quad G' \leftarrow G$
8: $\quad G \leftarrow FindNN(Data, M)$ //Find the nearest neighbors to M
9: $\quad M \leftarrow FindCenters(Data, M)$ //Find the new centers of the groups in G
10: $\quad i \leftarrow i + 1$
11: **until** $i == Iterations$ OR $G == G'$ //Stop after iterations or groups don't change
12: $Output: G, M$

3.2 Hierarchical

There are many agglomerative hierarchical algorithms, based on the notion of merging the closest pair and updating the data, treating the pair as a single data element. For example in Figure 3.1, the first pair to be merged is $(4, 30)$. This pair is now considered a single data item, and the next possible pair is merged, and so forth. The criterion by which the groups are merged distinguishes the agglomerative algorithm. In the generic agglomerative hierarchical algorithm in *Algorithm 3.2*, the *MergeGroup*() function contains the specific merging criterion, that operates on the proximity matrix A. A is updated to reflect the merged groups and the proximity between groups, and not just the (i, j) data item pairs. From the output of the merged pairs and the merging criterion value, a permutation of the data element leaves and the dendrogram hierarchy can be constructed.

Algorithm 3.2 *AgglomerativeHierarchy(Data, A)*

1: $N \leftarrow Size(Data)$
2: $\mathcal{G} \leftarrow Groups(Data)$ //One data element per group in $\mathcal{G}$
3: **repeat**
4:     $(\mathcal{G}', A') \leftarrow MergeGroup(\mathcal{G}, A)$ //Merge closest group
5:     *Output*: Merged Group Pair and Merge Criterion Value
6:     $(\mathcal{G}, A) \leftarrow (\mathcal{G}', A')$ //Update matrix and groups
7: **until** $|\mathcal{G}| == 1$

The simplest merging criterion, and of rather limited value, generates what is known as *single link* clustering. It merges two clusters based on the minimum dissimilarity between candidate groups, namely, the minimum nearest neighbor pair among all current groups. It has the effect of merging subgraphs that are the corresponding minimum spanning tree for successively larger subgraphs. Complete link clustering algorithm has as its merge criterion the furtherest neighbor criterion. In a graph theory sense, this criterion can be expressed as finding the minimum clique, where all pairs of dissimilarities (edges) need to be joined within a group, at each merge step. In single link, the graph analogy is expressed as only a single edge need be considered. Wards clustering merges groups based on minimizing the squared error between candidate groups, and thereby considers the minimum variance as the merging criterion instead of the proximity values directly.

3.2.1 Dendrograms and Heatmaps

In Figure 3.2 a generic dendrogram is depicted with two types of level selection. A cut at a level of 0.7 partitions the data into five clusters and an arbitrary cut represented by the dotted line, partitions the data into ten clusters. The arbitrary cut is just an example of the fact that one can make any cut to create a partition that represents a clustering, however, the straight

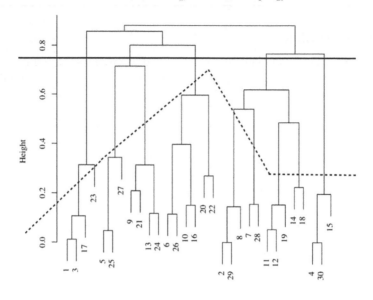

FIGURE 3.2: The solid horizontal bar across the dendrogram represents a cut or level in the tree that partitions the data into 5 clusters at a height of 0.75. The dotted line represents an arbitrary cut in the tree that creates 10 clusters.

line level cut at 0.7 is a partition consistent with the merge criterion which typically bounds the groups to similar extent within the feature space. The y axis represents the value of the merge criterion. This may be directly related to the measure such as in complete link and single link clustering or it may be a derived value that is quite distinct from the initial measure, such as in Wards clustering. The groups that are being merged, starting with the leaves, are not fixed in their position left to right. In fact, the ordering is a non-unique permutation of the data items such that the layout of the dendrogram does not cross lines and also the branches can easily be thought to rotate like a mobile. For example in the tree in Figure 3.2, given a slightly different permutation, data item 22 could be rotated along with its pair item 20 into position near the item pairs 13 and 24. (giiven that the merge of items 20 and 22 at the 0.6 level is the axis about which the rotation could occur). This makes the dendrogram sometimes misleading, in that sometimes it makes a group appear to be closer to another group than it might actually be. Its horizontal associations are not representative necessarily of the true structure of the data.

The first two dendrograms, Figures 3.1 and 3.2, in this chapter are con-

39 Cox2 Actives: 2D Fingerprint Data

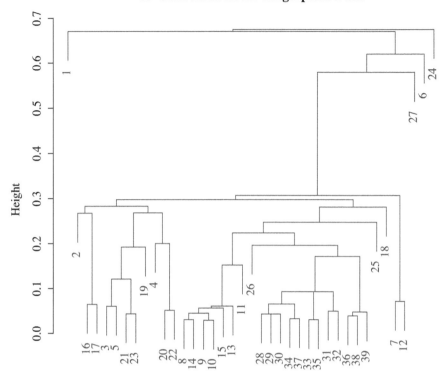

FIGURE 3.3: Single link hierarchy of 39 Cox-2 actives.

trived data used to display what dendrograms for these kinds of clustering algorithms typically look like. In Figures 3.3 and 3.4, we show the clustering of Cox-2 actives with the use of structural (2D) fingerprint data of the Cox-2 chemical compound ligands. The dissimilarity measure used in both is $1 - Tanimoto$. The difference between the nearest neighbor properties of single link and the farthest neighbor characteristics of complete link clustering can be noticed in the height at which groups are merged: merging of the complete link groups are typically at higher values. The tree topologies are significantly different upon simple inspection. In both of these dendrograms the height is the direct measure used to determine dissimilarity. With the same data and measure, Figure 3.5 shows a square error merge criterion called Wards clustering. The height is no longer in terms of the underlying measure, but in the merged criterion values. The same data set from Figure 3.5 can be plot-

39 Cox2 Actives: 2D Fingerprint Data

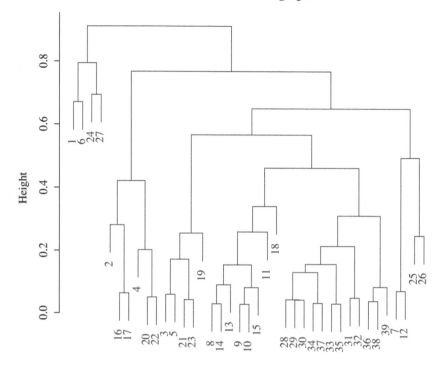

Leaves

FIGURE 3.4: Complete link hierarchy of 39 Cox-2 actives. Note how the topology of the tree has changed from the use of single link clustering hierarchy.

ted with a banner plot (see Figure 3.6). Banner plots provide visualization of hierarchical clustering results (e.g., Wards). The banner plot in Figure 3.6 plots the hierarchy such that the merge height is on the x-axis. The white bars extend to the merge height of the corresponding data item to its group. Groups can be distinguished by the long white bars. For example, item 39 is in a different group than item 2. Their merge height is at the top of the hierarchy and the white bar between them extends to the maximum height.

With small sets it is sometimes worthwhile to look at the dissimilarity matrix in the form of a heatmap, permuted to the rows and columns of one or two clustering algorithms. In Figure 3.7, the matrix rows and columns are permuted to the complete link dendrogram. The matrix is thus symmetric under this permutation. Figure 3.8 however uses Wards clustering for the

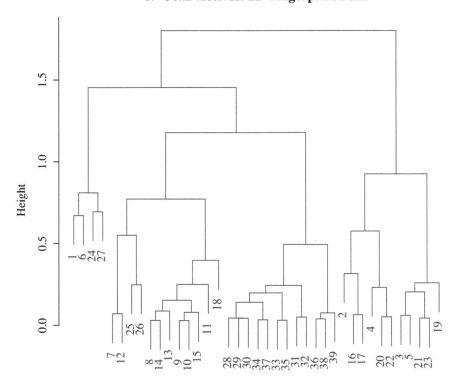

FIGURE 3.5: Wards hierarchy of 39 Cox-2 actives. Again, the topology differs from both Figures 3.3 and 3.4.

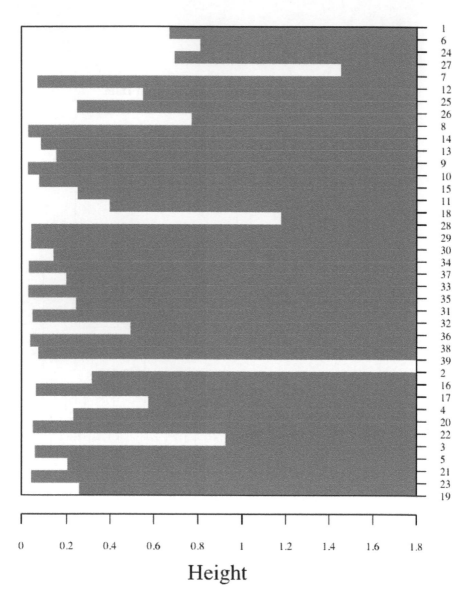

FIGURE 3.6: The Wards hierarchy of 39 Cox-2 actives from Figure 3.5 in banner plot form.

column permutation, and complete link for the permutation of the rows. Observe how some clusters are preserved at specific cuts, such as the group (22, 20, 4, 16, 17, 2).

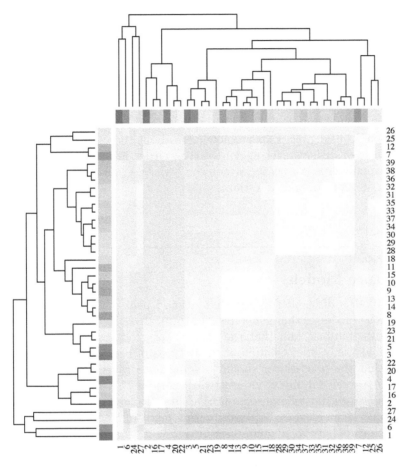

Complete Link Clustering

FIGURE 3.7: Complete link clustering heatmap.

Divisive hierarchical algorithms have as a basic premise finding a single division of the data that minimizes or maximizes some criterion, and subdividing the subsequent groups similarly. Splitting criterion can be simple or computationally complex and are very much like the types of splitting criterion used in recursive partitioning in supervised learning (regression or decision trees).

Algorithm 3.3 *DivisiveHierarchy(Data)*

  1: $N \leftarrow Size(Data)$

2: $\mathcal{G} \leftarrow SingleGroup(Data)$ //$|\mathcal{G}| = 1$, there is just 1 group in G
3: **repeat**
4:    $\mathcal{G} \leftarrow DivideGroups(\mathcal{G})$ //Divide groups in $\mathcal{G}$
5:    *Output*: Merged Group Pair and Merge Criterion Value
6: **until** $|\mathcal{G}| == N$

The recursive nature of divisive algorithms make them only less efficient at building an entire hierarchy in general than agglomerative algorithms. However, if only a portion of the hierarchy is created with some non-trivial stopping criterion, divisive methods can be fast. It is equally true that agglomerative methods can be stopped at a certain level without creating the entire hierarchy and can also be made equally fast if not faster. But some stopping criterion is needed for both methods, which if quantitative may slow the procedures such that the benefit is negligible. Expert knowledge about the data however may warrant the use of divisive over agglomerative, either in terms of efficiency or efficacy. Equally true is whether or not the divisive criteria are any better than merge criteria is an open question.

3.3   Mixture Models

Mixture models often presuppose that a good deal about the data are known as to the processes that generated the data, that the data follow some probability distributions, and that the number of groups are more or less known. We make only brief mention of these types of clustering algorithms, in the case that the practitioner should be confronted with such data, and because there are several important statistical notions that relate to clustering in general - a cluster follows some probability distribution. Figure 1.4 in chapter 1 would be an ideal set to cluster with a mixture model given that we assume the clusters have a gaussian distribution and we could estimate very roughly the four means and variances for the four clusters. One such method is known as the Expectation Maximization (EM) algorithm, in which we could do just that: supply the algorithm with four starting means and variances, and after so many iterations the algorithm would converge and approximate the true means and variances of the four clusters. If we supplied too few or too many mean variance pairs, or means and variances that were wildly off from the true mean and variances of the clusters, the results would likely be nonsensical.

3.4 Sampling

Very large sets of data tax computational resources. Thus, methods have been developed that, rather than inspect all of the data pairwise relationships necessarily, the methods instead sample pairwise relationships and form partitions of the entire set based on the sampling. This form of clustering is akin to Monte Carlo integration, where the notions of pseudo and quasi-random sampling come into play. It is also based on a simple fact that if data item A is very dissimilar to data item B, and data item C is very close to data item B, then it can be inferred that the data item A is also very dissimilar to data item C. If the dissimilarity relationships between A and B, and B and C have been computed, then there may be no need to compute the dissimilarity between A and C. Choose a random data item and determine all of the dissimilarity values of the remaining data items to it. Include those items that are within some near neighbor threshold value, let that be a cluster, and then exclude them from further consideration. Repeat this with the remaining data items, forming all of the subsequent clusters similarly. Since we have excluded the items of the first cluster without determining their dissimilarity to any of the other items, but for the random one chosen, there are $k \times (N - k - 1)$ dissimilarities that need not be computed, where k is the number of items in the exclusion region minus the random one chosen. For each iteration, the size of the cluster membership will likely vary, but in principle the number of dissimilarities computed is only $O(|G| \times N)$, where $|G|$ is the number of clusters, and not the $O(N^2)$ computation for generating all pairwise dissimilarities.

Algorithm 3.4 $SimpleLeader(Data, Threshold)$

1: $DataElement \leftarrow SampleRandom(Data)$ //Arbitrarily choose a data element from the data
2: $Data < -Data \setminus DataElement$ //Remove DataElement from Data
3: $Append(DataElement, C)$ //Append the DataElement to first cluster, C
4: **repeat**
5:    $C \leftarrow Append(FindNN(Data, Threshold, DataElement), C)$ //Append NNs to cluster, C
6:    $Append(C, \mathcal{G})$ //Append cluster C to set of clusters $\mathcal{G}$
7:    $Data < -Data \setminus C$ //Remove cluster elements from data
8:    $DataElement \leftarrow SampleRandom(Data)$ //Arbitrarily choose a data element from the data
9: **until** $|Data| == 0$
10: $Output : \mathcal{G}$

In another form of cluster sampling the entire pairwise dissimilarities are computed, but only a small fraction of the values are stored and operated on by another clustering sampling algorithm. Namely, if there is a dissimilarity threshold, where we know that we are not concerned with groupings outside that dissimilarity threshold, we can choose not to store those values nor operate on them during the clustering. Though all pairwise dissimilarities have been computed, the clustering sampling algorithm will likely proceed much faster given a much smaller set of dissimilarity values to operate on. One peculiarity of the exclusion region algorithms is that they tend to produce a great many singletons (clusters with only one member each). The singletons are of two types given this algorithm: Singletons that are not near any other data element given the threshold, and singletons that are within the threshold of some neighbor or neighbors, but have been left out of any exclusion region. These *false singletons* can be gathered up and distributed to an exclusion region cluster where they have a nearest neighbor within the threshold.

Algorithm 3.5 *ExclusionRegion(SparseMatrix)*

1: **repeat**
2:    $C \leftarrow FindLargestSetNN(SparseMatrix))$ //Assign largest row in matrix to cluster, C
3:    $SparseMatrix \leftarrow SparseMatrix \setminus C$ //Remove row and its elements from the matrix
4:    $Append(C, \mathcal{G})$ //Append cluster C to set of clusters $\mathcal{G}$
5: **until** $SparseMatrix == NULL$
6: $Output : \mathcal{G}$

3.5   Overlapping

There are many cases where groups or classes are confounded in that they overlap, given the measure or space that the data represents. Consider Figure 3.9. Assume that the data in this scatter plot represents the true four classes shown in terms of their point marker. These classes overlap, the boundaries shown with the convex hull about each class, yet our disjoint algorithms will partition these into four groups, ignoring the overlapping of the classes. If the disjoint constraint is allowed to lapse somewhat, we may be able to get overlapping clusters that better define the classes in terms of the data features and measures that describe the data in this overlapping fashion.

In some cases, experts in the field in question may prefer overlapping clusters from disjoint ones, where there is either ambiguity as to membership of some data items, or some data items could easily have several class memberships. Consider for example how it is sometimes uncertain what group an organism belongs to in a taxonomy. Rather than force a class decision, it may be of more use to represent the uncertainty in having the organism belong to more than one group. There are clustering algorithms that return results that do just that. In many cases, an algorithm that returns disjoint clusters can be modified to return overlapping clusters.

The generic exclusion region Algorithm 3.5 can be modified to produce overlapping clusters given false singletons, under the assumption that it is possible for false singletons to have nearest neighbors in more than one exclusion region. One could argue that it would be better to distribute the false singletons to that exclusion region with its nearest neighbor. This would indeed produce disjoint regions. However, with discrete measures, there can be ties in nearest neighbor values. This ambiguity can be used to produce overlapping clusters. Figure 3.10 displays the fraction of a dataset involved in overlaps verses the Tanimoto threshold applied. A comparision of three different fingerprints is made with the shortest key based fingerprint revealing the largest portion of the data set participating in overlapping groups.

Algorithm 3.6 *ExclusionRegionOverlapping(SparseMatrix)*

1: **repeat**
2:     $TiedRows \leftarrow FindLargestSetsNN(SparseMatrix))$ //Find largest row(s)
3:     $i = 1$
4:     **repeat**
5:         $C \leftarrow AssignClusters(TiedRows[i]))$ //Assign rows to cluster non-disjoint clusters
6:         $Append(C, \mathcal{G})$ //Append cluster C to set of clusters $\mathcal{G}$
7:         $i = i + 1$
8:     **until** $i == |TiedRows|$//Repeat until largest tied rows processed
9:     $SparseMatrix \leftarrow SparseMatrix \setminus TiedRows$ //Remove row and its elements from the matrix
10: **until** $SparseMatrix == NULL$
11: $F \leftarrow FindFalseSingletons(SingletonClusters(\mathcal{G}))$
12: $\mathcal{G}' \leftarrow DistributeFalseSingletons(F, \mathcal{G})$ //Distribute false singletons to any cluster with a NN
13: $Output : \mathcal{G}'$

Since overlapping clusters are a cover rather than a partition, any significant overlapping can be explosive combinatorially. If the groups are well-separated at the scale of the threshold, there will be very little ambiguity, and in practice in this instance most of the overlapping concerns a few false singletons, and perhaps a few overlapping groups of the same size.

3.6 Fuzzy

Fuzzy clustering is related to the principle of overlapping clusters, but unlike the simple overlapping clustering expressed in the last section, data elements have a degree of membership across those clusters that might contain them. It is sometimes useful to regard the degree of membership as a probability, but it need not be considered in those terms necessarily. Somewhat analogous to K-means, fuzzy clustering attempts to iteratively refine the degree or probability of membership to those groups that are most alike under some criterion. Thus, for example, at the end of the number of iterations, many data elements have very little or no degree of membership for most clusters, and only for a few clusters are their respective degrees of membership significant, or even exclusive to a single cluster. Data items can be assigned disjoint clusters given their respective highest final membership coefficient, or the coefficients can be used to assign membership to more than one cluster to form a cover rather than a partition.

Algorithm 3.7 $Fuzzy(Data, K, Threshold)$
1: $U \leftarrow FEmptyLists(K)$
2: $M' \leftarrow AssignMembership(Data, K)$ //Assign memberships
3: **repeat**
4:      $M \leftarrow M'$
5:      $U \leftarrow FindCenters(Data, M, K)$ //Find the cluster centroids
6:      $M' \leftarrow ComputeMembership(Data, U)$ //Compute memberships to current cluster
7: **until** $MembershipDifference(M, M') < Threshold$
8: $\mathcal{G} \leftarrow AssignGroups(M)$ //From final membership, M, assign groups to $\mathcal{G}$
9: $Output$: $\mathcal{G}$, M

Fuzzy clustering typically has a learning parameter that is somewhat arbitrary across a certain range, and even updated and changed over the span of the iterations. Tuning parameters such as these are also found in the following section, Self-Organizing clustering methods. They bring to these methods an amorphous almost artful feel in determining what initial and subsequent values to input to arrive at useful clusterings. These methods can converge fairly quickly and arrive at reasonable solutions efficiently, but they are not typically fast enough for very large scale clustering.

3.7   Self-Organizing

Self-Organizing clustering algorithms use either a preset number of clusters or nodes, or generate clusters (not necessarily predefined) where the clusters are adapted from the data. The main distinction of SO algorithms is this adjustment of many (possibly all) of the cluster nodes or a neighborhood of them to the data items individually, rather than calculating the centroids simply as a response to near neighbors as in K-means [158].

The most popular Self-Organizing algorithms heavily utilized in bioinformatics and to some extent also in drug design are SOM (Self-Organizing Maps) and SOTA (Self-Organizing Tree Algorithm). The specific algorithms for SOM and SOTA are discussed in Chapter 4 and Chapter 7 respectively.

3.8 Hybrids

The design of algorithms [124] is not just concerned with various forms of efficiency but also efficacy. This is especially true of clustering algorithms, where researchers are often faced with new and different data sets that present challenges for the common clustering algorithm software they may have at hand. In almost every field attempts are made at designing new clustering algorithms, often out of those portions of existing algorithms that have some property the designers have confidence in, success with, knowledge of, or simply that it will fit with some other aspect of another clustering algorithm. These attempts are not always successful, but that does not stop them from necessarily being published in the literature.

3.9 Glossary

agglomerative: grouping starting from individual data items (bottom up)

clustering hierarchy: a ranking of subset relationships that reflects the feature space of the objects, given a measure of similarity and a merging criterion. Sometimes referred to as representing a *topology* of those relationships.

disjoint partition: disjoint subsets of a set (no members in common), the union of which is the set.

divisive: splitting into groups recursively

false singleton: a member identified as an outlier which is within a threshold distance from a cluster member

fuzzy: an assignment of partial membership

random sampling: the random sampling of some distribution (uniform, normal, unknown, etc.) whether generated via some sequence on a computer such as pseudo or quasi-random sequence generation, or via some natural process.

Voronoi diagram: a diagram in which all line segments are equidistant between any two pairs of points in two dimensions which can be generalized to higher dimensions with hyper planes (*Voronoi tessellation*)

3.10   Exercises

1. Given the dendrogram in Figure 3.2, assume the data used to generate the dendrogram is two dimensional and in a Euclidean space. Draw the data points in a 2D rectangle as best you can, trying to represent the groupings. Pay close attention to the level at which groups are merged. Draw a convex hull around each of the 5 groups formed by the horizontal cut of the dendrogram with solid lines. Similarly, with dotted lines, draw the convex hulls around the 10 groups formed by the dotted line cut. Can you think of reasons why the dotted groups would be of use? What drawbacks do you see in this method? How do they compare with the groups found with the level cut? How might these different types of cuts impact the clustering validation?

2. The scatterplot in Figure 3.11 is a multidimensional scaling (MDS) of the data used for the Cox-2 hierarchies in Figures 3.3-3.5. Choose a non-trivial cut in any of the three dendrograms (single link, complete link, and Wards), and list the cluster elements by number next to the groups found in the scatter plot. Don't be surprised if some of your groupings are incorrect: remember the scatterplot is a 2D approximation of the relationships. Note, the axes labeled values are dimensionless with respect to the dissimilarity measure used to compute the hierarchies.

3. Find examples of overlapping groups (genetics, drugs, etc.) and define the features of the data that leads to the overlapping. Compare the extent of the overlapping.

4. Design a setup for a fuzzy clustering problem concerning human genetics and geographic location. Describe the features and the initialization of the problem. Discuss what results you might expect.

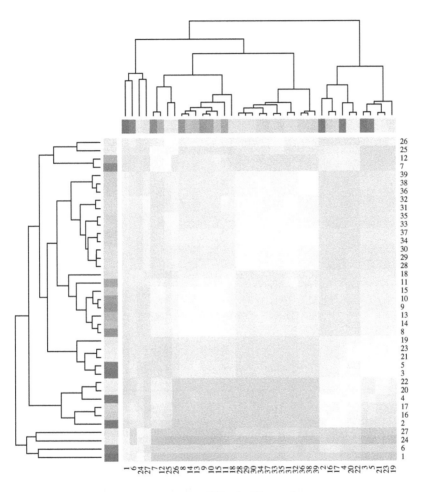

FIGURE 3.8: Wards and complete link clustering heatmap.

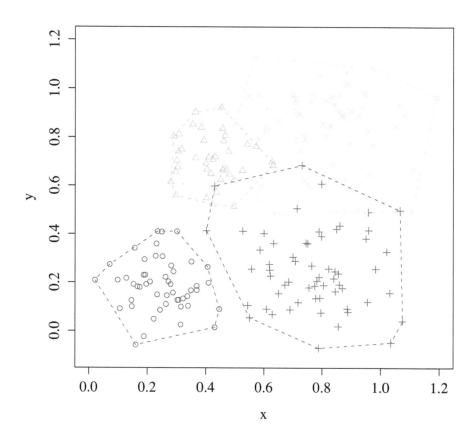

FIGURE 3.9: A contrived data set with four overlapping clusters, generated with random normal points with four centers and varying standard deviations, similar to Figure 1.5.

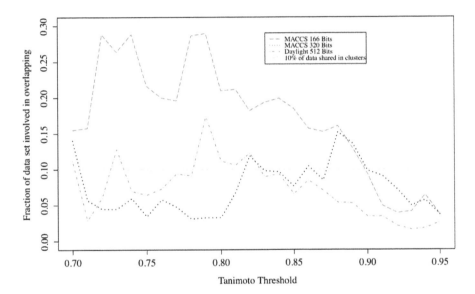

FIGURE 3.10: Clustering ambiguity in Taylors overlapping clustering given 380 HIV active compounds and three different fingerprint types.

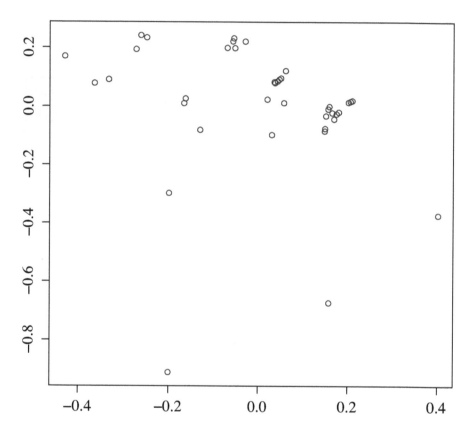

FIGURE 3.11: MDS plot of Cox-2 dissimilarity data

Chapter 4

Partitional Algorithms

Partitional clustering algorithms are quite popular in both bioinformatics and drug discovery. Variations of K-means continue to spring up in the literature, largely in an effort to tackle larger and larger data sets efficiently. Jarvis-Patrick was a popular method for clustering large scale data sets of binary fingerprints early on in cheminformatics, but has been supplanted largely by K-means-like and sampling algorithms. Spectral clustering, popular in the imaging community, has had some adherents for medium sized data sets in cheminformatics. Self-organizing maps of various forms have become popular in bioinformatics, though originally coming from the engineering literature. They have an analog computing flavor with numerous parameters that vary throughout the clustering process, but it can be shown that they are really quite similar to K-means in their overall behavior. Their popularity is in part due to the fact that they provide a convenient way in which to visualize the results as a part of the process and output.

4.1 K-Means

A basic form of the K-means algorithm shown in chapter 3 (Algorithm 3.1) can be modified at will seemingly (e.g., see [97, 49, 100, 66, 82, 128, 95] for various examples). Some are due to the types of data, others are for performance reasons or notions of attaining a more nearly optimal partition by careful selection of the starting centroids. The few rationales are presented here and examined in connection to efficacy and data type. In addition, though online clustering algorithms are important in fields such as cyber security and fraud detection of enormous streaming data sets, and are rarely needed in bioinformatics and drug discovery applications, they may be useful for clustering very large data sets. The algorithm starts by generating arbitrary centers within the feature space of the data. Such points can be either randomly generated or actual data points randomly selected from the data, either pseudo randomly or quasi-randomly chosen (e.g., Van der Corput sequence [52] scaled to the indices of the data).

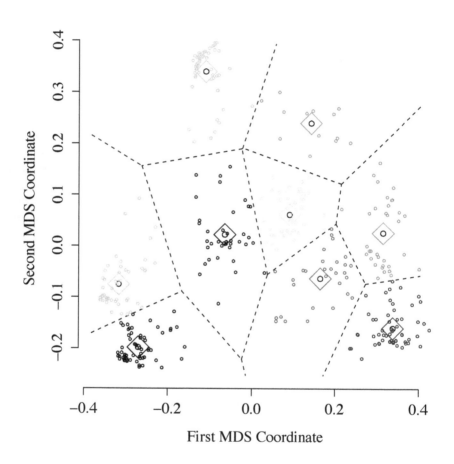

FIGURE 4.1: Simple *K*-means with two dimensional data set derived from MDS of 405 Benzodiazepine ligand data set. Each cluster's centroid is marked by a square. Note, centroids are not coincident necessarily with data points. Each cluster is bounded by its respective Voronoi region.

Algorithm 4.1 $Kmeans(Data, K, Iterations)$

1: $\mathcal{G} \leftarrow KEmptyLists()$
2: $M \leftarrow ArbCenters(Data, K)$ //Find K arbitrary centers
3: $\textbf{\textit{Axis}} \leftarrow \textbf{\textit{PrincipalAxis(M)}}$ //Find principal axis of centroids
4: $\boldsymbol{\mathcal{G}} \leftarrow \textbf{\textit{FindNN(Data, M, Axis)}}$ //Find the Nearest Neighbors to M
5: $M \leftarrow FindCenters(Data, M)$ //Find the centers of the groups in $\mathcal{G}$
6: $i \leftarrow 1$
7: **repeat**
8: $\quad \mathcal{G}' \leftarrow \mathcal{G}$
9: $\quad \textbf{\textit{Axis}} \leftarrow \textbf{\textit{PrincipalAxis(M)}}$ //Find principal axis of centroids
10: $\quad \boldsymbol{\mathcal{G}} \leftarrow \textbf{\textit{FindNN(Data, M, Axis)}}$ //Find the Nearest Neighbors to M
11: $\quad M \leftarrow FindCenters(Data, M)$ //Find the centers of the groups in $\mathcal{G}$
12: $\quad i \leftarrow i + 1$
13: **until** $i == Iterations$ OR $\mathcal{G} == \mathcal{G}'$
14: $Output: \mathcal{G}, M$

4.1.1 K-Medoid

The K-medoid algorithm is very much like K-means, but instead of finding centroids that may not be actual data items in K-means, K-medoids always uses actual data items as the groups' centers. The term *medoids* has the flavor of measure of central tendency like a median value, versus a mean, and the medoids or centers themselves are often called *centrotypes* or *exemplars*. From a statistical standpoint, this approach is also very much like choosing the median versus the mean of a distribution as a measure of central tendency as it is less effected by outliers and noise. In practice this is often a modest improvement over K-means results that can be especially useful when the data does indeed have outliers and noise.

Just as in K-means, an arbitrary set of K centrotypes can be chosen randomly, or more thoughtful initial data centrotypes can be chosen [82]. Rather than find centroids, K-medoid swaps each data point with each centrotype and calculates a measure (often the average or sum of distances) to determine if the new configuration is better under some cost function, and if so, whether therefore to swap the centrotype to the new data point. This is done repeatedly until no centrotypes are swapped. However, with discrete data, it is possible that arbitrary choices must be made that may cascade to other swaps.

Algorithm 4.2 $Kmedoid(Data, K, Iterations)$

1: $\mathcal{G} \leftarrow KEmptyLists()$
2: $M \leftarrow ArbCenters(Data, K)$ //Find K arbitrary centrotypes within Data
3: $\mathcal{G} \leftarrow FindNN(Data, M)$ //Find the Nearest Neighbors to M
4: $M \leftarrow FindCenters(Data, M)$ //Find the centers of the groups in $\mathcal{G}$
5: $i \leftarrow 1$
6: **repeat**

7:    $M' \leftarrow Swap(Data, M)$ //Perform swaps to determine M
8:    $\mathcal{G} \leftarrow FindNN(Data, M')$ //Find the centers of the groups in $\mathcal{G}$
9:    $i \leftarrow i + 1$
10: **until** $i == Iterations$ OR $M == M'$
11: *Output:* $\mathcal{G}$, M

4.1.2 K-Modes

K-modes algorithm is nearly identical to K-medoid, but for the fact that it operates on categorical data and the corresponding measures of (dis)similarity. Binary string data can be regarded as categorical data, bringing into play the various binary data measures. The centrotypes are swapped in and out again in connection with a cost function determined now by the categorical measures. In the context of binary strings, rather than the mean or medoid center, modal fingerprints (see Chapter 2 modal fingerprint definition) can be used as centers. The parameterization of the modal fingerprints can significantly alter the results, and, given the discrete nature of the fingerprints or categorical data, K-modes is even more likely to have odd swapping effects like K-medoid.

Algorithm 4.2 $Kmodes(Data, K, Iterations)$
1: $\mathcal{G} \leftarrow KEmptyLists()$
2: $M \leftarrow ArbCenters(Data, K)$ //Find K arbitrary centrotypes within Data
3: $\mathcal{G} \leftarrow FindNN(Data, M)$ //Find the Nearest Neighbors to M
4: $M \leftarrow FindModal(Data, M)$ //Find the centers of the groups in $\mathcal{G}$
5: $i \leftarrow 1$
6: **repeat**
7:     $M' \leftarrow Swap(Data, M)$ //Perform swaps to determine M
8:     $\mathcal{G} \leftarrow FindNN(Data, M')$ //Find the centers of the groups in $\mathcal{G}$
9:     $i \leftarrow i + 1$
10: **until** $i == Iterations$ OR $M == M'$
11: *Output:* $\mathcal{G}$, M

4.1.3   Online K-Means

Online K-means performs the K-means algorithm with two small twists. The closest centroid to an input data item is updated as the data are streamed in one by one. The updating of the centroids is no longer the mean of the nearest neighbors, but rather by some small fraction λ. Thus, if the data item is $\mathbf{x}_j$ and its closest centroid is $\mathbf{m}_i$, then the centroid is updated immediately via:

$$\mathbf{m}'_i = \mathbf{m}_i + \lambda \mathbf{x}_j. \tag{4.1}$$

This form of the algorithm is very much like Self-Organizing Maps (see Section 4.4.1). In fact, it differs very little from an online version of SOM. It is fairly easy to see that this impacts the stability of the results with respect to input order regardless of the data type.

Algorithm 4.4 $OnlineKmeans(DataStream, K, \lambda)$

1: $\mathcal{G} \leftarrow KEmptyLists()$
2: $M \leftarrow ArbCenters(Data, K)$ //Find K arbitrary centers
3: **repeat**
4:     $j \leftarrow FindNN(DataStream_i, M)$ //Find NN in M of newest data item
5:     $M \leftarrow UpdateCenter(M_j, DataStream_i, \lambda)$ //Update M_j with Equation 4.1 and store in M
6:     $\mathcal{G} \leftarrow UpdateGroup(j, DataStream_i)$ //Update $\mathcal{G}$ with corresponding new data item
7: **until** $DataStream == \emptyset$
8: $Output$: $\mathcal{G}$, M

4.2 Jarvis-Patrick

There are two basic forms of the Jarvis-Patrick algorithm based on constraints imposed on nearest neighbor tables generated to form the partition. The most basic constraint is the overall length of the nearest neighbor table to be considered, K. Namely, only those nearest neighbors up to K are to be considered in assembling the clusters. The next constrain is K_{min}, where $K_{min} \leq K$, and $K_{min}, K_{min} \geq 1$. At least K_{min} near neighbors must match between the any two respective K length near neighbor lists to be within the same cluster. Both of these constraints presents problems with ties in proximity leading to cluster ambiguity that will be considered in chapter 9.

Algorithm 4.4 below sketches the simple *Jarvis-Patrick(K, K_{min})* version. K near neighbors must be found for all rows in $FindNN(Data, K)$, requiring sorting of each row, $O(N^2 \log N)$ overall - or K calls to $min()$. Of course if $K > \log N$, this is of no help. In worst case, naively the algorithm must search an additional $O(KN^2)$ pairs. In practice, the interplay of K_{min} and K_{min} and the data make worst case performance unlikely. Creating and sorting the near neighbor lists is the bulk of the execution time. The algorithm has a near neighbor form of chaining akin to the minimum spanning tree. So on the one hand the algorithm is not constrained to hyper-spherical (exclusion region algorithms) or hyper-rectangular (recursive partitioning algorithms) regions as some algorithms produce, but it often has many singletons as the chaining is somewhat localized. The algorithm was often used for fairly large scale clustering in cheminformatics [120], FindNN(Data,K) is embarrassingly parallel, but the core portion of Jarvis-Patrick is not. Efforts to scale the algorithm to fit very large data sets through parallelization is therefore somewhat limited. However, medium sized near neighbor tables in the tens of thousands are quite manageable with Jarvis-Patrick. It is therefore a convenient algorithm to have in ones toolbox for data set of this size, if for no other reason than for comparison to other algorithm results.

To get a feel for how this algorithm works with a very small sample of real data, the SMILES strings of twelve compounds from a set of benzodiazepines were converted to a key-based fingerprints of 768 bits each and then compared with the Tanimoto similarity measure. The similarity measure values are in Table 4.1. Each row of columns one through twelve can be considered an unranked near neighbor table. Bold cells in the neighbor list rows denote the $K = 5$ nearest neighbors; cells with identical similarity scores are ties that lie at the $K = 5$ constraint; underlined cell numbers are those that are proximity ties that are within the $K = 5$ constraint. The precision of the comparison values is truncated to fit conveniently in the table, but those proximity values are exact rational numbers, and bold and underlined values are indeed ties in proximity.

The last column in the table contains the list of $K = 5$ nearest neighbor indices. The clusters generated by Jarvis-Patrick with $K = 5$ and $K_{min} = 3$ are $(1, 10, 11)$, $(2, 3, 4, 5, 6, 7, 8, 9)$, and the singleton cluster, (12). The reader can inspect the last column of the table and identify the $K_{min} = 3$ constraint that produces the three clusters. Note that index 12 is not in any of the $K = 5$ lists. Changing K_{min} to $K_{min} = 2$ produces the following clusters $(1, 3, 4, 5, 10, 11)$, $(2, 6, 7, 8, 9)$, and the singleton cluster, (12). This change moves data elements 3, 4, and 5 into the first cluster. The constraint is loosened, as now only two near neighbors need match between two rows to be considered in the same cluster. Again, by inspection with the $K = 5$ lists, the reader can assemble these clusters as well.

Algorithm 4.4 *Jarvis $-$ Patrick(Data, K, K_{min})*

1: $NNK \leftarrow FindNN(Data, K)$ //Create nearest neighbor table with K length constraint

2: $p \leftarrow 1$

3: **repeat**

4:     $i \leftarrow 1$

5:     $j \leftarrow i + 1$

6:     $C_p \leftarrow i$ //Add i to cluster C_p

7:     **repeat**

8:         **if** K_{min} criterion is met between (i, j) rows for all j in NNK **then**

9:             $C_p \leftarrow j$ //Add j to cluster C_p

10:          $RemoveRow(NNK, j)$ //Remove the j_{th} row from NNK

11:         **else**

12:             $j \leftarrow j + 1$

13:         **end if**

14:     **until** $j > RowLength(NNK)$

15:     $RemoveRow(NNK, i)$ //Remove the i_{th} (first) row from NNK

16:     $p \leftarrow p + 1$

17: **until** $NNK = NULL$

18: *Output*: C //Return set of clusters C

TABLE 4.1: Example of *Jarvis-Patrick*$(K = 5, K_{min} = 3)$

FPs	1	2	3	4	5	6	7	8	9	10	11	12	K = 5
1	1.000	0.452	0.472	0.472	0.455	0.432	0.460	0.452	0.472	0.723	0.691	0.331	(3,4,9,10,11)
2	0.452	1.000	0.717	0.717	0.701	0.941	0.861	0.845	0.813	0.300	0.295	0.377	(4,6,7,8,9)
3	0.472	0.717	1.000	0.960	0.915	0.681	0.730	0.732	0.752	0.301	0.296	0.325	(4,5,7,8,9)
4	0.472	0.717	0.960	1.000	0.952	0.681	0.730	0.732	0.752	0.301	0.296	0.342	(3,5,7,8,9)
5	0.455	0.701	0.915	0.952	1.000	0.694	0.743	0.716	0.735	0.299	0.295	0.339	(3,4,7,8,9)
6	0.432	0.941	0.681	0.681	0.694	1.000	0.865	0.798	0.786	0.287	0.283	0.371	(2,5,7,8,9)
7	0.460	0.861	0.730	0.730	0.743	0.865	1.000	0.861	0.846	0.315	0.310	0.338	(2,5,6,8,9)
8	0.452	0.845	0.732	0.732	0.716	0.798	0.861	1.000	0.960	0.3	0.295	0.333	(2,4,6,7,9)
9	0.472	0.813	0.752	0.752	0.735	0.786	0.846	0.960	1.000	0.321	0.316	0.358	(2,4,6,7,8)
10	0.723	0.300	0.301	0.301	0.299	0.287	0.315	0.300	0.321	1.000	0.948	0.273	(1,4,7,9,11)
11	0.691	0.295	0.296	0.296	0.295	0.283	0.310	0.295	0.316	0.948	1.000	0.278	(1,4,7,9,10)
12	0.331	0.377	0.325	0.342	0.339	0.371	0.338	0.333	0.358	0.273	0.278	1.000	(2,4,5,6,9)

The second common Jarvis-Patrick variant removes the length of the nearest neighbor list, K, and supplants the K_{min} constraint with a percentage, P_{min}, the minimum percentage of nearest neighbors that need to match between any two nearest neighbors to be considered within the same cluster.

Algorithm 4.5 *Jarvis−Patrick(Data, P_{min}, Threshold)*

1: $NNK \leftarrow FindNN(Data, Threshold)$ //Create nearest neighbor table at a (dis)similarity threshold
2: $p \leftarrow 1$
3: **repeat**
4:     $i \leftarrow 1$
5:     $j \leftarrow i + 1$
6:     $C_p \leftarrow i$ //Add i to cluster C_p
7:     $K_{min} \leftarrow min(length(NNK(i)), length(NNK(j)))$ //K_{min} constraint min of row lengths of i and j
8:     **repeat**
9:         **if** K_{min} criterion is met between (i, j) rows for all j in NNK **then**
10:             $C_p \leftarrow j$ //Add j to cluster C_p
11:             $RemoveRow(NNK, j)$ //Remove the j_{th} row from NNK
12:         **else**
13:             $j \leftarrow j + 1$
14:         **end if**
15:     **until** $j > RowLength(NNK)$
16:     $RemoveRow(NNK, i)$ //Remove the i_{th} (first) row from NNK
17:     $p \leftarrow p + 1$
18: **until** $NNK = NULL$
19: *Output*: C //Return set of clusters C

4.3   Spectral Clustering

Spectral clustering was introduced in the late 1990s in conjunction with image processing. For example, clustering of pixels is an important part of image segmentation, where pixels may be vectors of RGB colors or the seven spectral bands common in satellite imagery. Such problems can be relatively large, though the dimension is typically small, in the hundreds of thousands or millions of pixels, so spectral clustering is typically done on small images or subsets of larger images. Luckily, computing eigenvalues and eigenvectors of fairly large matrices is possible, making spectral clustering of modestly large data sets in the thousands of data items in higher dimensions more reasonable. Therefore, spectral clustering has begun to be used in cheminformatics applications, with several groups now exploring its efficacy with respect to various data types within the early drug discovery process [18, 58].

There are typically two methods, one [103] very much akin to PCA or MDS that reduces the dimension of the data by computing k eigenvectors of the Laplacian matrix from the similarity matrix, A:

$$L = I - D^{-\frac{1}{2}} A D^{-\frac{1}{2}}, \tag{4.2}$$

where the diagonal matrix has entries $\sum_{j=1}^{N} A_{ij}$, for the ith entry. The choice of the dimension is simply some $k \ll N$, where the k eigenvectors act as the columns of the data features. These are very much like loadings or linear factors. These transformed data are in turn clustered by any clustering algorithm, though most commonly by K-means.

The other more interesting method [122] again computes the Laplacian matrix, but recursively partitions the the data based on the first and second eigenvalues, building a hierarchy from the top down.

4.4 Self-Organizing Maps

Self-Organizing Maps is specifically a non-linear projection of an N-dimensional space onto a special form of a 2-dimensional surface. The algorithm can be thought of as a cross between K-means and 2-dimensional elastic or force directed MDS methods. The surface however has a special property in that it is a torus - the edges of the grid that comprises the SOM wrap around. Thus, clusters that form along the edge are related closely to the clusters on the opposite edge. SOM is sometimes referred to as *topologically ordered maps* because they try to preserve the proximity relationships from the original space of the problem to a simpler space or topology, and not necessarily because the results are in the form of a curious topological object such as the torus. Grid centers on the toroidal projection of the data are *stretched* (in keeping with the elastic and topological metaphor) towards single data points, but, akin to K-means, just grid centers near the point. This process is done iteratively, such that all of the points are run through this process up to a number of specified iterations. The algorithm is thus parameterized by how much the grid centers are moved by each point; what the size of the neighborhood defines closeness; the number of iterations. "Grid" here is broadly defined to refer to a simple uniform tiling. So for example the grid need not be square but can be hexagonal. SOM also samples the original space much like K-means and does a good job at creating dense clusters and placing them within grid regions appropriately.

SOM is particularly sensitive to input parameters and can form unfortunate kinks in the map under certain conditions, where the learning is localized. Additionally, SOM is sensitive to input order - a change in the input order may cause a change in the learning that is additive, leading to a different, albeit similar, result, and though the result may be similar it may take some effort to show that it is similar.

In our description of the algorithms we will refer to the grid points as *Nodes*. Nodes are adjusted by moving them linearly some fraction towards each individual data point based on the learning rate function, $\lambda()$, and the neighborhood function. The number of nodes that make such a change under consideration decreases as given by the neighborhood function, $H()$, as the number of iterations increases. $H()$ is an indicator function whether or not a node moves. The total number of nodes is k, representing in the end k clusters, though it is possible that a node may be empty, in practice this is rare, so the choice of k is very much like the choice of k for any partitional clustering method. We use i here to represent a node that is under consideration, from the total set that may be under consideration at a specific iteration. We can think of the movement of the nodes as a very simple linear adjustment:

$$N' = N + \alpha\,(\mathbf{x} - N) \tag{4.3}$$

where α is a single linear coefficient between zero and one, moving the node N some fraction towards the data item $\mathbf{x}$. The full expression for moving a node towards a single data item is,

$$N_i(Iter + 1) = N_i(Iter) + \alpha\,(\mathbf{x}_j - N_i()) \tag{4.4}$$

where $\alpha = \lambda(Iter)H(Iter)$, and

- N_i is the ith tree node in question,

- $Iter$ is the number of iterations, the learning rate,

- λ is a function that is inversely proportional to the iteration,

- $H()$ is the neighborhood function, such that as the iteration increases the nodes that change in response to the jth input data, $\mathbf{x}_j$, come from an ever smaller neighborhood of nodes closest to $\mathbf{x}_i$, as $Iter$ increases.

The **Algorithm 4.3** $SOM(Data, K, Iter, \lambda(), H())$

```
 1: InitializeKNodes() //Randomly assign coordinates for the nodes
 2: ParameterInitialization() //Iterations, λ(), H()
 3: Iter ← 1
 4: n ← SizeData()
 5: repeat
 6:    j ← 1
 7:    repeat
 8:       i ← 1
 9:       repeat
10:          Node_i ← UpdateNode(Node_i, x_j) //Update a node position given
             x_j
11:          i ← i + 1
```

12:      **until** $i > K$
13:      $j \leftarrow j + 1$
14:   **until** $j > n$
15:   $Iter \leftarrow Iter + 1$
16: **until** $Iter > Stop$
17: $Output : SOM$

4.5   Glossary

global optima: the maximum or minimal optimal point or region within the entire search space

iterative gradient optimization: minimizing or maximizing search along gradients to find an optima(local or global).

local optima: within a search space, there maybe local optimal regions that are not optimal over the entire search space

nearest neighbor: neighbor in a set of comparisons which is the closest.

single axis boundarizing: a projection of the feature space onto the first principal component in order to bound a search space, typically for nearest neighbor searching.

4.6   Exercises

1. Create a set of sample data sets that demonstrates the impact of the input parameters of K (and K_{min} in the case of Jarvis-Patrick) on the clustering results for each of the algorithms discussed in this chapter. Focus the dicussion on the changes in the cluster membership for different limits of K, (e.g., $K_{min} \ll K$ and $K_{min} = K$ for Jarvis-Patrick).

2. Create a rank order of the algorithms in this chapter, listing the algorithms you would expect to be the most efficient (or have the capability to handle the largest data sets). Include in your discussion what issues you might foresee in cluster validity (are the results *reasonable* for each of the algorithms described here).

3. Perform a literature search and find a recent paper that applies SOM for clustering in a drug design or bioinformatics application. Discuss

the advantages and disadvantages of the SOM approach for the particular application described. Propose a different algorithm for the analysis and discuss the advantages and disadvantages of its application for this particular reference.

4. Compare the performance of K-means and Spectral Clustering for large data sets, in terms of time and space. Pay particular attention to how they would perform in practice.

Chapter 5

Cluster Sampling Algorithms

Criteria for forming a partition with purely partitioning algorithms operate on an iterative adjustment of a feasible or temporary partition, taking into account all of the data items. In K-means, a typically randomly chosen set of K centroids are picked to start the iteration. Another similar set of algorithms approaches the forming of partitions, by randomly choosing a single starting centroid, and forming a cluster by gathering all of the other data items that are within a pre-defined similarity threshold. The remaining data items - those excluded from the cluster - are then treated similarly, repeating the process until all data items are in some cluster. The clusters are more directly sampled as it were from the data, and there is no pre-selected number of clusters as there is in, say, K-means. The primary benefit is efficiency, either in the case of the leader algorithms by not having to store a proximity matrix, nor having to calculate all (dis)similarities, but rather $O(KN)$ (dis)similarities, where $K = |\mathcal{C}|$ is the eventual number of clusters; or, in the case of the exclusion region algorithms, having only to store a sparse matrix of the similarities, where the clustering step amounts to an $O(KN)$ lookup of some K number of clusters. The assumption in terms of efficacy is that these algorithms sample the density of the data in the feature space, and thereby sufficiently identify the groups. The leader algorithms lend themselves as a last resort to the very largest data sets or huge online data streams. The exclusion region algorithms tend to be more effective as they select clusters from what are clearly dense regions of the feature space, determined by the fact that they select the largest near neighbor entry in the near neighbor table at a specific threshold, and are not quite as subject to problems of random selection, but these algorithms suffer from the cost of generating all-pairs of similarities.

5.1 Leader Algorithms

In K-means, the algorithm is typically initialized with K randomly chosen points in the feature space, or randomly selected data items. Such random selection can be performed with either pseudo or quasi-randomly generated sequences, both in an effort to *spread* the selection of the K starting centroids somewhat uniformly about the feature space that contains the data. In this way, via the iterative process, the adjusted centroids will more likely find regions of density, such that those regions are well defined and separate from one another. Compare this notion to the leader algorithm, where the initialization of the algorithm starts with a single random choice of a data item and an exclusion threshold to be used to generate clusters. If the data has a tendency to cluster, then there are dense regions in the feature space, and the initial random point will more likely be in one of these dense regions. But, nevertheless, there is a small but likely significant probability that it will not be. It may be an outlier for example. This is especially true of high dimensional spaces, where it is often the case that some data items will lie at the periphery of the feature space. Though again small, there is also a significant probability that the randomly selected data item will lie between clusters and may draw into the first cluster what would otherwise be some members of several other clusters. Once the first cluster is defined and excluded, the question arises how to best continue the random selection of the remaining points to obviate some of the problems of the choice of the first random selection.

Like K-means, one would like to be assured that each new random sample from the remaining, yet unclustered data, is spread again somewhat uniformly throughout the feature space. Variations on the leader algorithm concern just that: strategies for choosing the next random sample point in the remaining unclustered data. A simple first pass at this is to use information from what has been calculated so far. At each iteration a $1 \times (N - |\mathcal{C}_i|)$ set of dissimilarities has just been calculated, where $\mathcal{C}_i$ is the set of clusters up to the ith iteration, and $|\mathcal{C}'|$ is the number of data items so far clustered. Those dissimilarities that lie out side the range of the threshold are all of the remaining data items that have not been clustered, and from which the next random data item will be chosen. Which one? One method is to choose the furthest from the most recent random data item, requiring a small computational hit for finding the maximum in the list. This process may have a tendency to pick outliers about the edge of the features space. Another method is to select arbitrarily any data item outside the exclusion region, with only a small probability that the data item will be very near an exclusion region, at least at the beginning of the algorithm when there are just a few clusters. Another method is to choose, if possible, a point that lies 2 times the radius of any previous exclusion region. This latter method is most akin to the K-means uniform random sampling, as it successively chooses points that are spread throughout the data, and it has the flavor of a quasi-random sampling, method known as Poisson disk sampling for Monte Carlo integration, still an active area of research in computer graphics [29, 123, 51]. This method however requires storing and processing all previous dissimilarities, so it is both slower and uses significantly more memory, but it may be more effective as it will sample the feature space density of the data with a Monte Carlo integration-like process. The type of space is defined by the metric or measure and therefore its *exclusion region* can be quite different than our usual notion of Euclidean space and distance. This is especially noticeable with the discrete data and measures, where the space and the regions are really on a lattice. Lastly, there are other methods that store and process even more dissimilarities [12, 88], and though less efficient than the simple methods mentioned above, are nevertheless efficient compared to the exclusion region algorithms where all $O(N^2)$ dissimilarities need to be calculated, and other partitioning algorithms.

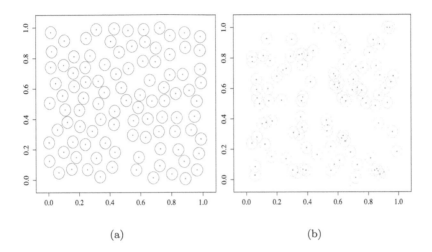

(a)   (b)

FIGURE 5.1: Figure a) Poisson disk sampling of the unit square with radius 0.04; Figure b) Uniform pseudo random sampling of the unit square, showing disks of 0.04 radii.

To get an intuitive sense of the sampling in both leader and exclusion region algorithms, Figure 5.1a shows a quasi-random point set of 100 points created by Poisson Disk sampling - no point can be within $2r$ of one another, where r is the radius of the disk. The field of points covers the space more uniformly, even though the points themselves were generated by a pseudo random generator, excluding points that fit within the radius of any previously sampled point. One-hundred pseudo random points were generated for Figure 5.1b with circles of the same radius used to generate the Poisson sampling in 5.1a. It is easy to see that even though the points are drawn from a uniform pseudo random number generator ("runif()" in the R statistical computing language)[133], the sampling is far from uniform. This example is just in 2D Euclidean space, but it generalizes to more dimensions and to other dissimilarity measures. How the sampling picks regions to be excluded sequentially also has an impact on the final groups chosen, and hence the rationale for instance of picking the next sample point outside of $2r$, where r is the exclusion region threshold radius of any previous cluster in the leader algorithms, to help sample the density much like a quasi-Monte Carlo integration method.

In the leader algorithm 5.1 below, the additional concerns of sampling above and beyond that found in the *SimpleLeader* Algorithm 3.4 are shown as an generic function call *SamplingMethod(Data)*, that is a catchall for any of the above variants of selecting the next sample centroid.

Algorithm 5.1 *Leader(DataSet, Threshold)*

1: $DataElement \leftarrow SampleRandom(Data)$ //Arbitrarily choose a data element from the data
2: $Data \leftarrow DataSet \setminus DataElement$ //Remove DataElement from Data
3: $Append(DataElement, C)$ //Append the DataElement to first cluster
4: **repeat**
5:    $G \leftarrow Append(FindNN(Data, Threshold), C)$ //Append NN to cluster
6:    $Append(C, G)$ //Append cluster C to set of clusters G
7:    $Data \leftarrow Data \setminus C$ //Remove cluster elements from data
8:    $\mathbf{DataElement} \leftarrow \mathbf{SamplingMethod(Data)}$ //Choose new centroid
9: **until** $|Data| == 0$
10: $Output : G$

Online versions of this algorithm can also be used for vast data sets, where only a subset of the data are used in the iterative process, then additional data are added either to the now existing clusters, or new clusters are created. This provides additional speed up as not all of the data are considered at once, but the loss of the additional dissimilarity comparisons given all of the data, lessens the amount of possibly useful information that can be used for the sampling procedure. The online versions rely even more on the notion that the feature space density of the data can be sampled randomly with considerable effect. This is somewhat data dependent, but almost surely there is an additional drop off in efficacy as a result of more chance of obtaining a poor sampling.

A slightly more complex variant of this algorithm is called *Leader-Follower*. In this algorithm, the data items are associated with the random sample data item, but that initial centroid is adjusted by the inclusion of the data point by some small factor, λ towards the data item, such that subsequent data items are gathered as it were with respect to the new centroid that is very likely not a data item at all. New centroids are randomly selected from the data items when a data item does not fit within any region of an already existing centroid. This algorithm is very closely related to online K-means. Leader and Leader-Follower algorithms also lend themselves naturally to online clustering (also known as *competitive learning* such as online SOM and other neural net approaches).

5.2 Taylor-Butina Algorithm

Exclusion region algorithms are similar to leader algorithms but require that all pair-wise dissimilarities have been computed in order to generate a near-neighbor table at the specified threshold. Taylor and Butina arrived at a similar algorithm separately [132, 22]. Their method is on the surface extremely simple. A near neighbor table at a threshold is a sparse matrix with N number of rows. The empty rows are data items that have no near neighbors and are thus singleton clusters. These can be removed first. Successively, the row with the most neighbors is considered a cluster and it is removed, and all of its members are removed if they are in any other row of the matrix. This process is repeated with the newly formed sparse matrix. Each new cluster is smaller for each new iteration. The iterations stop when the matrix contains all empty near neighbor tables. However, this leaves out what are known as false singletons: data items that are within a threshold of some other cluster members, but not a cluster member themselves. The false singletons can be assigned to the clusters with the nearest neighbor cluster members.

Algorithm 5.2 $TaylorButina(SparseMatrix, Threshold)$

1: $CollectSingletons(SparseMatrix)$ //Remove singletons.
2: **repeat**
3:    $C \leftarrow FindLargestSetNN(SparseMatrix))$ //Assign largest row in matrix to cluster
4:    $SparseMatrix \leftarrow SparseMatrix \setminus C$ //Remove row and its elements from the matrix
5:    $Append(C, \mathcal{G})$ //Append cluster C to set of clusters $\mathcal{G}$
6: **until** $SparseMatrix == NULL$
7: $\mathcal{G} \leftarrow AssignFalseSingletons(\mathcal{G})$//Assign false singletons to clusters with nearest neighbors
8: $Output : \mathcal{G}$

Figure 5.2 provides a 2D visualization of the Taylor-Butina algorithm run at a threshold of 0.05. The centroids are depicted as the solid circles, the threshold is defined by the large circles, diamonds display the singletons and the xs symbolize the false singletons. Figure 5.3 and Figure 5.4 provide visualizations of Taylor-Butina algorithm in 2D run at two different threshold values. In these figures the cluster rank order is described by number labels for each of the clusters. The thresholds 0.1 and 0.2 are dissimilarity thresholds. Hence, the cluster groupings in Figure 5.4 are larger than those groups depicted in Figure 5.3. Similarly, each group is typically more spread out. The larger exclusion regions also show more overlapping (a behavior discussed in Chapter 9). Note that the centrotypes (actual centroids in this instance) have a spacing that is more in line with the quasi-random sampling. The exclusion region circles overlap because instead of the sampling radius being $2r$ as in the poisson disk sampling, the exclusion region centroids radii are at least r.

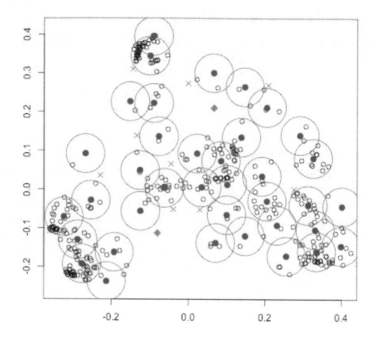

FIGURE 5.2: Taylor-Butina visualization in 2D at a threshold of 0.05. Centroids are depicted in solid circles, cluster members are small black circles, diamonds are singletons, x symbolizes a false singleton, exclusion regions are defined by the large circles.

This is just a constant factor in the sampling and has no impact in terms of the discrepancy of the points.

One of the supposed drawbacks of the Jarvis-Patrick and Taylor-Butina algorithms is that they tend to have a lot of singletons. Exclusion region algorithms in general have little or no "chaining" as one would find in other types of clustering, most notably at the extreme, single link clustering or minimum spanning tree. Outliers therefore remain as such clusters. As a result, an additional chaining-like feature is added and a significant amount of what would otherwise be included in the set of "true" singletons is removed.

There are on occasion several tie breaking decision issues with this algorithm that will be discussed in greater detail in Chapter 9, but are briefly mentioned here. The largest near neighbor row may have more than one candidate. This tie in the number of near neighbors can be broken arbitrarily, or some measure of density can be applied to each of the tied rows to determine which row is the densest. An example of such is: choose the row with the minimum sum of all cluster members to the representative (centroid or centrotype). This is simple and fast as it is simply the sum of the row entries. Even with various tie breaking schemes this algorithm is typically not invariant to input order. Not all ties might be resolved analytically - there may even be minimum sum of distances ties for example. This problem is especially true when the number of cluster members gets small. However, if the tied clusters do not overlap, and thereby change the remaining clusters, the resulting clusters will be the same, only output in a different order. If there are a set of tied candidate rows for the next cluster and a least one pair do overlap, the choice of which row is considered a cluster can significantly alter some of the clusters that come thereafter.

One of the nice features of this algorithm is that upon completion, the representative for each cluster is known and can be output as the first member of the cluster. Also, the sparse matrices, though typically only a constant fraction of the total $O(N^2)$ size, can often be as little as one percent of the full matrix in practice. This makes for a very fast algorithm for large data sets once the sparse matrix has been created. Generating very large proximity matrices is this algorithm's drawback for very large data sets. Even the sparse matrices can get very large, especially if the data set is huge.

5.3 Glossary

cluster representative: the centrotype

cluster sampling: clustering includes some form of sampling to arrive at clusters

Poisson disk sampling: a form of quasi-random sampling used in graphics formed by pseudo random points and exclusion region rejection

similarity threshold: a level that is set above which two objects being compared are considered similar

sparse matrix: a thresholded matrix

true singleton: outside any exclusion region or cluster member by the threshold dissimilarity

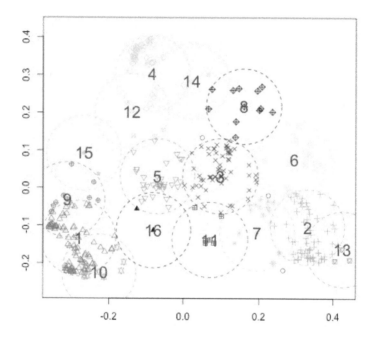

FIGURE 5.3: Taylor-Butina visualization in 2D at a threshold of 0.1. The order of cluster selection is labeled by numbers. Each number represents a different cluster.

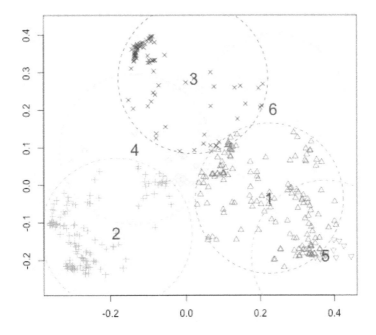

FIGURE 5.4: Taylor-Butina visualization in 2D at a threshold of 0.2. Each cluster is represented by a point marker and the cluster ordering is indicated by the numbers 1 − 6.

5.4   Exercises

1. Review Figures 5.3 and 5.4 that provide two different threshold visualizations for the Taylor-Butina algorithm. What might you expect in the visualization if the threshold value is increased to 0.3? Describe the concepts of cluster membership, singletons and false singletons as they apply to the visualizations.

2. Suppose you have a data set of highly similar compounds by a generic similarity measure. Describe what a visualization similar to Figure 5.3 or Figure 5.4 might look like for this kind of data set. What if the initial data set contained members that were highly unique?

3. Search for clustered data sets on-line which report singletons. Discuss which algorithms were used and what assessment, if any, can be made of the pre-clustered data.

4. Why are the centroids in Figure 5.2, Figure 5.3 and Figure 5.4 actually more like the sampling Figure 5.1a versus Figure 5.1b?

Chapter 6

Hierarchical Algorithms

Hierarchical clustering algorithms go one step further than providing a simple partition, and create a relationship of partitions that we can regard as a topology of the data items within a feature space, given a corresponding measure of similarity of that feature space. We need not concern ourselves with the mathematics of topological spaces, but use the term more loosely as simply a way of navigating the data organized by the set of nested partitions represented by the respective hierarchy.

In this chapter we delve more deeply into the specific hierarchical algorithms, focusing on the common, efficient, and more generally effective methods in the fields of bioinformatics and drug discovery, rather than exploring the full compendium of possible algorithms, many of which are simply variants on those offered here.

Hierarchical algorithms in general span those that create significant chaining such as single link clustering to those where clusters have very little if any chaining - and can be regarded as producing near hyper-spherical (broadly defined, or nearly metaphorical under some similarity measures) groups, such as those found in complete link clustering. This property, that we will call loosely, chaining to dense clusters, has its direct analogue in the spectrum of graph paths, k-connected components [77], to maximal cliques (complete link clustering). And this is not restricted to symmetric similarity relationships as we will see in Chapter 8, where asymmetric hierarchical clustering algorithm analogues use directed connectivity, starting with strongly connected components.

6.1 Agglomerative

Algorithm 6.1 is a modification of the Algorithm 3.2 for Agglomerative clustering methods with the MergeGroup function highlighted and the addition of G'. The Merging criterion for three agglomerative algorithms are defined in Equation 6.1, Equation 6.2 and Equation 6.3.

Algorithm 6.1 *AgglomerativeHierarchy(Data)*

1: $N \leftarrow Size(Data)$

2: $\mathcal{G} \leftarrow Groups(Data)$ //One data element per group in $\mathcal{G}$
3: **repeat**
4:     $\mathcal{G}' \leftarrow MergeGroup(\mathcal{G})$ //**Merge closest group**
5:     *Output*: Merged Group Pair and Merge Criterion Value
6:     $\mathcal{G} \leftarrow \mathcal{G}'$ //Update Groups
7: **until** $|\mathcal{G}| == 1'$

6.1.1   Reciprocal Nearest Neighbors Class of Algorithms

6.1.1.1   Complete Link

The complete link merging criteria in Equation 6.1 searches for the maximum dissimilarity between a newly formed cluster and an existing cluster.

$$G(w, (u, v)) = 0.5((G[w, u]) + (G[w, v]) - |G[w, u] - G[w, v]|) \qquad (6.1)$$

u, v represent a newly formed cluster and w represents an existing cluster [32, 105]. This merging criteria, by focusing on maximum dissimilarity, avoids the chaining problems of single link algorithms which focus on the minimum dissimilarity between clusters. However, the focus on maximum dissimilarity can put too much emphasis on the outliers for a given dataset. If the natural groupings in a dataset include true outliers, complete link algorithms will focus too strongly on trying to force outliers into a clustering group when they really should remain as outliers.

6.1.1.2   Group Average

The Group Average merging criteria in Equation 6.2 is true to its name by including the arithmetic averages of dissimilarity and taking into account the sizes of clusters.

$$G(w, (u, v)) = \frac{|u|}{|u| + |w|}(G[w, u]) + \frac{|v|}{|v| + |w|}(G[w, v]) \qquad (6.2)$$

u, v represents a newly formed cluster and w an existing cluster, where $|u|$, $|v|$ and $|w|$ are the cardinality of u, v and w respectively [90]. In Figure 6.1a a dendogram displays the results from a structural clustering of a set of known JNK3 actives at a level selection of 0.34 with Group Average clustering. In Figure 6.1b the centroids for the first five resulting clusters reading the dendogram from left to right are depicted. The cluster membership size for each cluster is displayed above the centroids in a text field. The 2D structures for each centroid are displayed as a graphic. Notice at this clustering level the 2D chemical diversity of these active compounds is apparent.

If one drills down to level 0.4 on the Wards dendogram depicted in Figure 6.2, the number of clusters increases from five clusters (largest has forty five members) and three singletons to seven clusters (largest has twenty nine members) and five singletons.

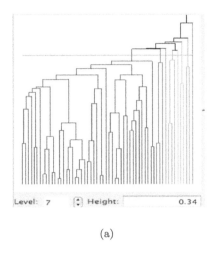

(a)

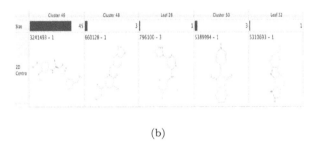

(b)

FIGURE 6.1: Some resulting structure clusters for a set of JNK3 active ligands clustered with Group Average at a level of 0.34 [48, 4, 104].

6.1.1.3 Wards

The Wards merging criteria in Equation 6.3

$$G(w,(u,v)) = \frac{|u|+|w|}{|u|+|v|+|w|}(G[w,u]) + \frac{|v|+|w|}{|v|+|u|+|w|}(G[w,v])$$
$$- \frac{|w|}{|u|+|v|+|w|}(G[u,v]) \qquad (6.3)$$

u, v represents a newly formed cluster and w an existing cluster, where $|u|$, $|v|$ and $|w|$ are the cardinality of u, v and w [144].

In Figure 6.3, the same set of JNK3 active ligands Figure 6.1 and Figure 6.2 are clustered in 2D with Wards hierarchical clustering. Four clusters result at the displayed selection level. Figure 6.4 represents the same clustering, again with Wards as in Figure 6.3, but with a level selection of one cutoff level lower

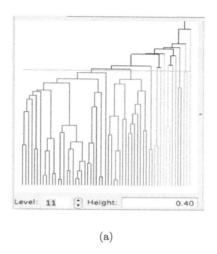

(a)

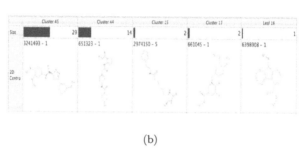

(b)

FIGURE 6.2: Some resulting structure clusters for the same set of JNK3 active ligands clustered with Group Average in Figure 6.1, but at a threshold level of 0.4 [4, 48, 104]. Cluster centroid for the first five nodes in the hierarchy are displayed with the cluster memberships listed above each centroid.

in the hierarchy. The number of clusters increases to five, and the centroids are displayed in Figure 6.4b. Three clusters remain the same and one cluster is split into two, resulting in five clusters displayed by the dendrogram in Figure 6.4a.

6.1.2 Others

There are many other forms of agglomerative hierarchical algorithms. The most common is single link clustering, which has such significant chaining that it is rarely useful for many applications. Others have properties such that they can not be made more efficient by **RNN**, so have minimal utility for data set sizes commonly analyzed in drug discovery. Still other hierarchical algorithms

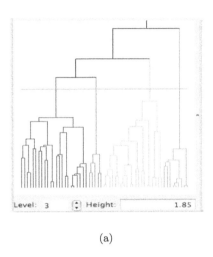

(a)

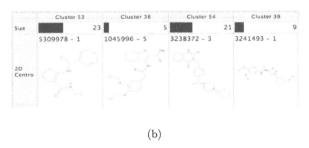

(b)

FIGURE 6.3: Some resulting structure clusters for a set of JNK3 active ligands [48, 4, 104] clustered with Wards. Cluster centroid for the four clusters that result from the level selection are displayed with the cluster memberships listed above each centroid.

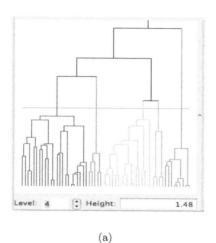

(a)

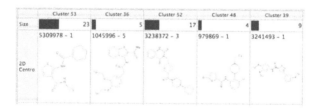

(b)

FIGURE 6.4: Some resulting structure clusters for a set of JNK3 active ligands [48, 4, 104] clustered with Wards. Same data set as in Figure 6.3, centroids represent cluster nodes for a level cutoff one level lower in the hierarchy than in Figure 6.3. Cluster centroid for the first five nodes in the hierarchy are displayed with the cluster memberships listed above each centroid.

have merging criterion such that they can only be used with metric measures. Other forms utilizing either weighted merging criteria or centroid merging criteria can fail to be montonic. Monotonicity is a property of the hierarchy consistent with ultrameticity.

$$d[(u, v)] \leq d[(w, (u, v)]$$ (6.4)

u and v are either groups or members. d is the merging criteria and w is another group or member. If Equation 6.4 does not hold, the clustering is not monotonic and crossovers can occur in the hierarchy.

6.2 Divisive

This chapter so far has only considered agglomerative algorithms, but there are methods which are hierarchical and divisive. Divisive algorithms begin with all data set members in one cluster and then divide the data from the top-down, recursively into smaller groups. These algorithms have as their advantage a more global consideration of the data than agglomerative methods. The disadvantage of divisive algorithms are that they can be computationally more expensive if the entire hierarchy has to be created. Divisive algorithms can be used in drug discovery or bioinformatics applications if the entire hierarchy is not built. The requirement that level selection occur at a high level in the dendrogram, increases the efficiency of these algorithms to be linear to data set size, but the challenge is devising meaningful stopping criteria akin to level selection, such that the resulting groups have validity [5].

6.3 Glossary

chaining: where an individual cluster is spread out in feature space versus spherical or near-spherical-like clusters

crossover: condition where the merging criterion value of a child group is greater than its parent group within the hierarchy

merging criterion: a quantitative expression used in an algorithm to join or merge groups or clusters

monotonicity: the property of the entire hierarchy such that there are no crossovers, conforming with Equation 6.4

reciprocal nearest neighbor chain: successive nearest neighbors where the last nearest neighbor in the chain matches the first member of the chain

splitting criteria: method used in an algorithm to divide a group or cluster

topology: properties which are conserved under deformation or transformation

---

6.4   Exercises

1. Download a gene expression dataset [48]. Cluster the data with the **R** statistical package [133] using each of the hierarchical clustering methods mentioned in this chapter. Comment on the hierarchies that result. Cluster first on genes and then cluster on experiments. Comment on your findings.

2. Download the structures and activities for a set of ligands which have shown activity both in an HTS assay and a secondary screening assay for a particular receptor, e.g., Focal Kinase Enzyme (FAK) [48]. Use FOSS (free and open source) software, e.g., CDK [23], to generate properties or structure fingerprints for the data set. Cluster the data hierarchically (as was done in Exercise 6.1). Comment on the dedrograms that result. Modify the input order of the input data set, re-run the analysis. Do any of the hierarchical algorithms discussed in this chapter appear input order dependent? If input order dependencies are observed, suggest any ideas you have as to why this might be the case.

3. Draw a dendrogram for a agglomerative algorithm and include an example of a crossover.

Chapter 7

Hybrid Algorithms

There is no real restraint to creating new clustering methods by way of drawing on basic components of other methods. There may be numerous motivations for such hybrid methods such as efficiency in time or space, or the developers may have good reason to believe that the new method is very effective for a certain type of data. Methods for bi-clustering are a good case in point, where there is a very specific clustering problem that is data driven. In this chapter, we will examine several newer methods that draw on both the partitional and hierarchical methodologies to create clustering methods that are used either in bioinformatics or drug discovery domains. The methods shown here of course are not exhaustive but they cover the broad notions of combining hierarchical and partitional methods in thoughtful and intriguing ways.

7.1 Self-Organizing Tree Algorithm

The self-organizing tree algorithm, commonly referred to as (**SOTA**)[70], has the rather unique property of building a hierarchy in an organic fashion, that is somewhat unlike traditional hierarchical clustering methods. Agglomerative hierarchical algorithms generate either a rooted tree or a forest of trees at a specific threshold, and level cuts of these trees create partitions that are clusters. Divisive hierarchies typically start at a single root and create a tree whose leaves may stop a non-trivial a partition, though again, level selection is used to create clustering partitions. SOTA is a divisive methodology of growing a tree from the top down, but the leaves are the final partition, and the splitting criterion is based on the SOM method. As part of the criterion, a measure is generated, the *Resource*, R_i, for the ith node in the tree, the sum of which over the generated tree can conveniently be used to validate the resulting clusters. The Resource sum is in fact a stopping criterion once a preset threshold is reached.

$$R_i = \sum_{j \in N_i} \frac{(\mathbf{x}_j - N_i)}{|N_i|}, \qquad (7.1)$$

where $|N_i|$ is the cardinality of the ith node in the tree (the number of data items associated with the ith node), and $(\mathbf{x}_j - N_i)$ is the distance between the jth data item and the N_i node as before in the Chapter 4 discussion of SOM. As the number of nodes grows with each round of iteration (adjusting nodes given each data item at time, over all nodes), called an *Epoch*, the total sum for all R_i up to that point is checked against the threshold to see if the tree need grow further.

In Figure 7.1, the SOTA-generated leaf nodes can be scattered throughout the levels of the tree, portrayed in the figure as solid circle endpoints within the tree.

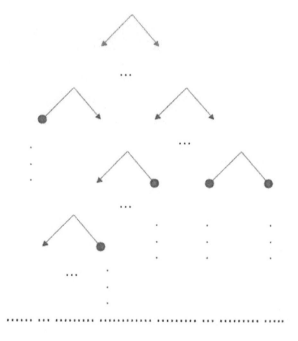

FIGURE 7.1: SOTA-generated tree example. The root of the tree is chosen typically as the mean or some measure of central tendency of the data. SOTA typically generates an unbalanced tree, where leaves representing clusters (solid circles can occur at various levels in the tree). The final partition are the groups formed by the leaves.

The expression for updating the internal position of the nodes, and which data items will be associated with which nodes, is much the same as in SOM, but now the learning rate function λ is fixed to just a few values, as the neighborhood function only contains a few possible neighbors: the closest external node, the sibling node if it is external, and the parent node. λ is greatest for the closest node, and the closest node therefore moves the most in response to $\mathbf{x}_i$.

$$N_i(Iter + 1) = N_i(Iter) + \lambda(Iter)H(Iter)(\mathbf{x}_j - N_i(Iter)) \qquad (7.2)$$

The algorithm starts out by assigning the mean of the data to the root node and two external siblings, and the first Epoch (or pass through the data) begins.

Algorithm 7.1 $SOTA(Data, \lambda(), H(), Threshold)$

1: $NodeInitialization(Data)$ //Find the mean of the data and assign it to the three nodes
2: $ParameterInitialization()$ //Threshold, $\lambda()$, $H()$
3: $n \leftarrow SizeData()$
4: $Iter \leftarrow 1$
5: **repeat**
6:   $j \leftarrow 1$
7:   **repeat**
8:     $i \leftarrow 1$
9:     $Rsum \leftarrow 0$
10:     **repeat**
11:       $Node_i \leftarrow UpdateNodes(Node_i, \mathbf{x_j})$ //Update a node positions given $\mathbf{x_j}$
12:       $Rsum \leftarrow Rsum + R_i$
13:       $i \leftarrow i + 1$
14:     **until** $i > |Nodes|$
15:     **if** $Rsum < Threshold$ **then**
16:       $BreakToOutput$
17:     **end if**
18:     $AddNodes()$
19:     $j \leftarrow j + 1$
20:   **until** $j > n$
21:   $Iter \leftarrow Iter + 1$
22: **until** $Iter > Stop$
23: $Output : SOTA$

7.2 Divisive Hierarchical K-Means

Divisive hierarchical K-means is a relatively simple hybrid algorithm, combining K-means and divisive clustering. Instead of the usual divisive splitting criterion, K-means is used as a splitting criterion (termed "Bisecting K-means" [128]). This makes for a fast algorithm as the K-means step is fast in general, using just two centroids (2-means) per group to be split. (In principle, the algorithm can be generalized to generate a general tree rather than strictly a binary hierarchy, by using $K > 2$. Determining the best K at each level however would be troublesome.) This algorithm was designed originally for large document searching, but has been applied more broadly since in drug discovery. The choice of each of the two initial centroids can vary, but most simply these can be chosen randomly, the problematic effects of K-means with respect to outliers notwithstanding.

Algorithm 7.2 *DivisiveHierarchical−K−means(Data)*

1: $N \leftarrow Size(Data)$
2: $\mathcal{G} \leftarrow SingleGroup(Data)$ $//|\mathcal{G}| = 1$, there is just 1 group in $\mathcal{G}$
3: **repeat**
4: $\mathcal{G} \leftarrow DivideGroupsK − means(\mathcal{G})$ //Divide groups, $\mathcal{G}$, at each level with K-means
5: *Output*: Merged Group Pair and Merge Criterion Value
6: **until** $|\mathcal{G}| == N$

7.3 Exclusion Region Hierarchies

An exclusion region clustering algorithm generates a partition at an exclusion threshold. Successive thresholds can be used to build a general rooted tree, but more typically a forest of trees whose roots start at a specific starting threshold [140]. The representatives or centrotypes form the successive nodes of the tree in an agglomerative fashion. This form of hierarchy can produce overlapping clusters, thus creating a general pyramidal clustering hierarchy [2, 7, 8].

The simple exclusion region Algorithm 3.5 in Chapter 3 can be modified to iterate over a sequence of successive thresholds, where the thresholds are increasingly larger—growing towards the root of the tree and allowing merging at each level. This algorithm requires considerable subroutine support, the details of which are glossed over. Figure 7.2 is a general tree example of an exclusion region generated hierarchy. The trees may be rooted or a forest of trees. The leaves represent a partition of the data at the last tier. The algorithm takes a sparse matrix at the stopping threshold and successively clusters at larger and larger dissimilarity thresholds until a threshold less than or up to the threshold at which the sparse matrix was create is reached.

Algorithm 7.4 *ExclusionRegion(SparseMatrix, StartThreshold, StopThreshold, Increment)*

1: $Step \leftarrow Increment$
2: $Threshold \leftarrow StartThreshold$
3: $Stop \leftarrow StopThreshold$
4: **repeat**
5: **repeat**

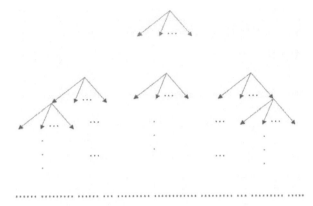

FIGURE 7.2: General tree example of an exclusion region generated hierarchy. Such trees may be rooted (top tier) or a forest (second row tiers) of trees. Leaves represent a partition of the data at the last tier (bottom row of dots represent the partitioned data).

6:      $SparseMatrix' \leftarrow SparseMatrix$ //Save copy of current SparseMatrix

7:      $C \leftarrow FindLargestSetNN(SparseMatrix))$ //Assign largest row index

8:      $SparseMatrix < -SparseMatrix \backslash C$ //Remove row and its elements from the matrix

9:      $Append(\mathcal{G}, C)$ //Append cluster C to set of clusters $\mathcal{G}$

10:      $Append(M, Centrotype)$//Append the centrotype of C to the list of centrotypes, M

11:      **until** $SparseMatrix == NULL$

12:      $SparseMatrix \leftarrow ResetMatrix(SparseMatrix', M)$//Collect all M_{ij} values from $SparseMatrix$

13:      $Output : M, \mathcal{G}$

14:      $Threshold \leftarrow Threshold + Increment$//Increment threshold for next layer

15: **until** $Threshold = Stop$

16: $AssembleHierarchy$ //Successive output of M and $\mathcal{G}$ form forest of hierarchy

7.4   Biclustering

Biclustering, sometimes called subspace clustering, co-clustering, or block selection, can be thought of as trying to perform feature selection in both dimensions: what features are important, and what data items are important. Approaches to the problem were first offered by Hartigan [64]. Nearly three decades later, approaches to the problem were first applied to gene expression data [27]. To find an optimal, global solution to this problem requires 2^{N+P} combinations of data items and features. Such a problem is NP-hard and clearly requires heuristics. As portrayed in Figure 1.7 the goal is to design an optimizing criterion that permutes both dimensions (e.g., genes and experiments) such that they align into subset blocks.

A naive approach is to simply cluster both dimensions as can be seen in Figure 3.7 or 3.8. It is evident in these examples that there are apparent blocks of possible interest. However, the specific permutations of the data by formation of the dendrogram in either dimension is not unique. There may be better more salient block features than revealed by these permutations. The application may also dictate how to optimize a block. There are four basic types of optimization goals:

1. Constant values along the rows

2. Constant values along the columns

3. Constant values over the entire block

4. A coherence of values over a block

Using Figure 1.7, the naive method of hierarchical clustering of both dimensions reveals blocks with near constant values for both rows and columns. (The original matrix in that example was of course symmetric, so there is a greater chance that there will be constant value blocks.) There are now a great many algorithms designed with these various goals in mind, each with features and shortcomings. For a survey on methods commonly used with gene expression data see [77]. A broader, more recent survey can be found in [87].

7.5   Glossary

hybrid algorithm: an algorithm that is a combination of algorithms that could be applied independently

pyramidal clustering: overlapping hierarchical clustering

SOTA epoch: one pass through all of the data

submatrix: a subset of contiguous rows and contiguous columns

---

7.6   Exercises

1. Discuss the advantages of applying the SOTA algorithm over utilizing SOM. Discuss any drawbacks.

2. Compare a hierarchical clustering algorithm to SOTA in terms of efficiency and efficacy.

3. Design a level selection technique for the exclusion region hierarchy.

4. Design a hybrid algorithm.

5. The permutation of a dendrogram layout of the leaves is not unique. There are several methods to permute the leaves such that there is a left-right arrangement that specifies the tightest merged clusters are to one side or the other. In **R** [133] there are two hierarchical clustering routines, "hclust()" and "agnes" (from the library "cluster"). They each have an associated plotting method, "plclust()" and "plot.agnes()" respectively. Both generate the same hierarchical clustering algorithms such as Wards, complete link, etc. Use these two routines and their plotting routines on a small gene expression data set along with the "heatmap()" function to produce different dendrograms and heatmaps. Use the same clustering algorithm (e.g., Wards) to create two heatmaps, one with "hclust()" and "plclust()" and one with "agnes()" and "plot.agnes." Inspect the results for all four types of biclustering optimization goals. Does the left-right permutation of "plclust()" reveal more types of blocks than can be found with "plot.agnes()". Use several different algorithms and inspect the results. Use different algorithms for each dimension.

Chapter 8

Asymmetry

Comparisons of objects with different numbers of nodes, length of strings, size or shapes creates a possible breakdown in the meaning or utility of a symmetric distance or similarity measure, like the Euclidean distance or the Tanimoto similarity. A very simple example from the social sciences regards clustering a larger group of individuals into a number of social cliques. Individuals rank their feelings or perceptions of one another, say, on a scale from 1 to 10. It is easy to see how a proximity matrix of such values is asymmetric since the pairwise values between individuals may differ - Ida likes Joey (the $(Ida, Joey)$ entry), but Joey just tolerates Ida (the $(Joey, Ida)$ entry). Such data can then be clustered to find groups of like-minded individuals (social cliques). Another simple example involves shapes. Two objects of the same shape, two triangles, might be quite "similar" to an observer, but a symmetric measure would measure them as quite dissimilar, if the two triangles are quite different in size. Thus, any asymmetric measure is one in which the order of the comparison of two objects may result in the a measure having different values. In Figure 8.1, there are two objects (in (a) and (b)) being compared. A similarity measure would find these objects to be quite "different" if their size difference is included in the similarity comparison. However, the fact that the triangle in Figure 8.1b can "fit into" the triangle in Figure 8.1a may be an interesting way to group these objects together. An asymmetric measure would have a high value for the case of Figure 8.1b being compared with Figure 8.1a and an low value for the case of the object in Figure 8.1a "fitting" into the triangle depicted in Figure 8.1b.

Comparisons of chemical structures in 2D or 3D are ubiquitous operations. Similarity searching and clustering of these compound descriptions are common methods in numerous early drug discovery applications. A standard technique is to encode chemical structures in the form of a binary string or fingerprint and compare different compounds by comparing their respective fingerprints with the Tanimoto similarity coefficient. Similarly, conformational shapes of a single compound or numerous compounds can be compared via a Tanimoto similarity coefficient with the use of shape descriptors defined by various methods, including Gaussian mixtures [56] and quasi-Monte Carlo integration based shape fingerprints [99] among others. Similarity coefficients have an inherent bias due to the simple fact that small molecular structures can nevertheless have a considerable size difference. Coefficients can be designed or composed to help alleviate such bias [35, 46], and one such method

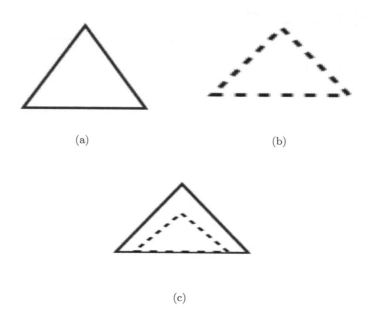

(a)   (b)

(c)

FIGURE 8.1: Figure 8.1a depicts a large triangle, Figure 8.1b depicts a small triangle. Figure 8.1c shows that the triangle from Figure 8.1b can fit into the triangle in Figure 8.1a, but if you try the reverse, only a portion of Figure 8.1a will fit into the triangle of Figure 8.1b.

is to use asymmetric measures to incorporate the bias. A common asymmetric measure used in a variety of fields (social sciences, biology, etc. [138]) is the Tversky similarity coefficient. The Tversky is like a parameterized version of the Tanimoto coefficient. Asymmetric measures can then be used to bias the comparison between two chemical structures. From a 2D, structural, chemical graph standpoint, if the substructure of a larger molecule is very much like a smaller molecule, the smaller molecule "fits in" the larger molecule and therefore has a larger similarity, whereas the larger molecule does not "fit in" to the smaller molecule in any strong sense, so it has a smaller relative similarity.

The Taylor-Butina asymmetric clustering results are displayed for a set of conformers for JNK3 active ligands clustered with an xray structure (2EXC) of a bound JNK3 ligand. The cluster representatives for the largest clusters are displayed in Figure 8.2b. Figure 8.2c displays the first few cluster members of the largest cluster (note 2EXC is the second member of the cluster and that the centroid of the cluster shown as the first depiction has a high Tversky score).

In Figure 8.2a, results for a Taylor-Butina asymmetric clustering of three dimensional molecular conformations are shown in a banded layout (akin to

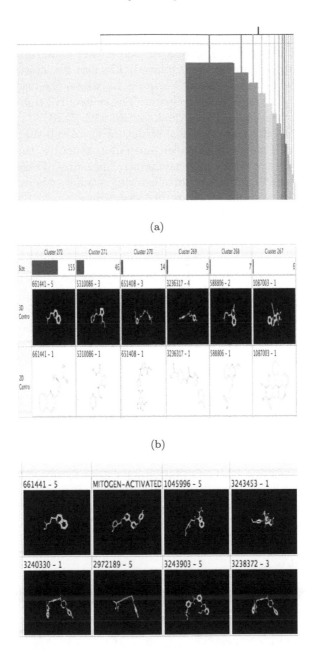

FIGURE 8.2: The Taylor-Butina asymmetric clustering results are displayed for a set of conformers for JNK3 active ligands clustered with an xray structure (2EXC) of a bound JNK3 ligand. The cluster representatives for the largest clusters are displayed in Figure 8.2b. Figure 8.2c displays the first few cluster members of the largest cluster, note 2EXC is the second member of the cluster and that the centroid of the cluster shown as the first depiction has a high Tversky score.

a hierarchy), rank ordered from left to right according to cluster size. Conformations of JNK3 active ligands are clustered with an x-ray structure (2EXC) of a bound JNK3 ligand. In Figure 8.2b, the cluster conformation representatives for the largest clusters are displayed. The first few cluster members of the largest cluster are displayed in Figure 8.2c, where the crystal structure, 2EXC, is the second member of the cluster. The centroid of this cluster, shown as the first depiction, has a high Tversky score with 2EXC.

Likewise in 3D, if a large chemical structure has a sub-shape that is very similar to the entire shape of a smaller molecular structure, these two shapes will have one of the two very similar asymmetric measures. These measures can then in turn be used in similarity or clustering studies, where the clustering is performed using asymmetric clustering algorithms.

Sequence alignments is an area of research in bioinformatics in which symmetric measures are not always the most effective comparison tool. Aligning sequences of different lengths and character share similar issues with the size/shape discussion of chemical structures described above. A subset of a sequence may exactly match a subsection of a larger sequence and knowledge of that 'similarity' in the case of genes or protein sequences should be reflected in the similarity scores in the analysis. Edit distance, a common measure applied in sequence comparisons, is discussed below. Edit distance is another asymmetric measure.

8.1   Measures

The social science example above is a form of empirical asymmetric data, but there are numerous methods that can be used to transform symmetric measures into asymmetric measures if the data warrants such comparisons. Below we discuss two common asymmetric measures used in bioinformatics and drug discovery.

8.1.1   Tversky

The Tversky measure is very much like the Tanimoto measure. Using the conventions in Chapter 2, the Tversky can be stated as follows:

$$s = \frac{c}{\alpha a + \beta b + c},$$ (8.1)

where α and β are continuous valued parameters. In order to get a value range in $[0, 1]$, and to relate the Tversky directly to the Tanimoto, one can constrain α and β so that $\alpha = 2 - \beta$. In this way, when $\alpha = \beta = 1$, the Tversky becomes the Tanimoto. The asymmetry results from swapping α and β in generating the similarity between (i, j) and (j, i). When one of the parameters is near zero the resulting similarity is strongly asymmetric, and when both parameters are near one, the resulting similarity has very little asymmetry.

8.1.2 Edit Distance

Edit distance is a commonly used measure in sequence alignment [11, 94]. Given two sequences the edit distance is the least costly transformation, namely the sum of the least number of deletions, insertions or substitutions necessary to transform one sequence into the other. For example, the edit distance to transform the string *EDIT* into *TEXT* is greater than the edit distance required to transform the string *NEXT* into *TEXT*, since the latter requires only a letter exchange and the transformation for the former requires more than one step (e.g., a letter exchange, a deletion and an insertion). There are potentially many possible transformations, the edit distance is the transformation which requires the least costly number of steps.

$$D_{ab} = \min \sum_{j=1}^{N} c_i(j) + c_x(j) + c_d(j), \tag{8.2}$$

Where D_{ab} is the edit distance between two sequences a and b, $c_i(j)$ is the cost of the insertion steps in the transformation, $c_x(j)$ is the cost of the exchange steps and $c_d(j)$ is the cost of the deletion steps for a given transformation, j, in the set of N all possible transformations.

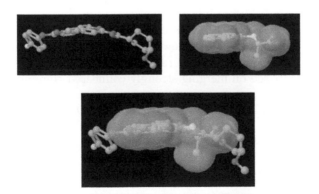

FIGURE 8.3: A Tversky similarity comparison between an xray structure (2EXC) of a compound with known activity to JNK3 (a kinase implicated in neuronal apoptosis) and a conformer of a ligand found to have activity to JNK3 in an screening assay [78, 48, 4, 139]. The overlay in 3D of the two structures shows a subshape match in Tversky space for these two JNK3 active ligands.

8.2   Algorithms

8.2.1   Strongly Connected Components Agglomerative Hierarchical Algorithm

Strongly connected components (SCC) is a directed graph algorithm that finds components in a graph if they admit to being cycles. More formally, a strongly connected component in the graph is a vertex set, where there is a directed path from vertex u to vertex v for all u and v in the vertex set. If the graph is weighted, the SCC algorithm can be used recursively to first find the smallest strongly connected components with the sum of the least weighted edges, and successively merging components with the sum of the least weights until there is just one strongly connected component representing all of the vertices in the graph. This process builds an agglomerative hierarchy from single data items through components of ever increasing size. Much like complete link and single link can be considered in terms of graph models (finding maximal cliques, and finding minimum spanning tree subgraphs respectively). The SCC agglomerative hierarchical algorithm is a directed graph approach that lies somewhere in between furthest neighbor behavior of complete link clustering and the chaining behavior of single link approach. Its closest analogue to undirected graphs is bi-connected components (there are at least two paths between u and v in the vertex set in an undirected graph). (Note, there are other forms of connectivity: tri-connected, k-connected, that have also been used in clustering [67, 119].)

An efficient algorithm for the use of SCC to create an agglomerative hierarchy is due to Tarjan [131]. To see how the method works however we show it here in the form of a simple agglomerative hierarchical algorithm, where the merge criterion is based on merging strongly connected components with the least edge weights.

Algorithm 8.1 *SCC AgglomerativeHierarchy(Data)*

1: $N \leftarrow Size(Data)$
2: $\mathcal{G} \leftarrow Groups(Data)$ //One data element per group in $\mathcal{G}$
3: **repeat**
4:    $\mathcal{G} \leftarrow MergeWeightedSCC(\mathcal{G})$ //Merge closest group based on SCC edge weights
5:    *Output*: Merged Group Pair and Merge Criterion Value //Value is the weight of the last edge added to create the SCC
6: **until** $|\mathcal{G}| == 1$

8.2.2   Modified Taylor-Butina Algorithm

The simplest asymmetric version of Taylor-Butina can be represented by the simple exclusion region algorithm, where the input matrix is a sparse asymmetric matrix. Here, the largest nearest neighbor table, whether in the row or column space, that is left in the matrix at each iteration is considered the next cluster. Setting up the asymmetric sparse matrix data structure is a bit more complex than the simple nearest neighbor table of the symmetric form of the algorithm, and the manipulation of that matrix to remove all of those members in the exclusion region also requires searching in both the row and column space of the matrix. The algorithm is therefore roughly twice as slow as the symmetric version, providing twice as much space is used to store both the row and column space. Smaller memory usage can be arranged, but only at the expense of the search time.

Algorithm 8.2 *AsymmetricExclusionRegion(AsymSparseMatrix)*

1: **repeat**
2:    $C \leftarrow FindLargestSetNN(AsymSparseMatrix))$ //Assign largest row (or column) in matrix to cluster
3:    $SparseMatrix \leftarrow SparseMatrix \setminus C$ //Remove row (or column) and its elements from the matrix
4:    $Append(C, \mathcal{G})$ //Append cluster C to set of clusters $\mathcal{G}$
5: **until** $SparseMatrix == NULL$
6: $Output : \mathcal{G}$

In a bit of irony, the above asymmetric clustering algorithm with the use of SAM (sequence alignment and modeling system with the use of hidden Markov models), akin to edit distance in generation of asymmetric associations, was utilized for bank customer retention segmentation clustering [110]. Clearly, the asymmetry inherent in genetic sequence data relationships could utilize the same technique.

8.3   Glossary

asymmetric matrix: a square matrix where the (i, j) element does not necessarily equal (j, i)

homology model: a prediction of a 3D protein structure for a novel protein sequence that shares sequence homology with a known protein structure

similarity searching: an $m x n$ matrix of similarities where $m \ll n$

skew symmetric: the average difference between the (i, j)th element and the (j, i)th element in an asymmetric matrix

8.4   Exercises

1. A Tversky similarity comparison between an xray structure (2EXC) of a compound with known activity to JNK3(a kinase implicated in neuronal apoptosis) and a conformer of a ligand found to have activity to JNK3 in an screening assay [78, 48, 4]. The overlay in 3D of the two structures shows a subshape match in Tversky space for these two JNK3 active ligands. Compare the shape of the 3D conformers displayed in Figure 8.3. If you maximize α and minimize β, would the Tversky value of the 3D conformer in Figure 8.3a into Figure 8.3b be closer to 0 or 1? For Figure 8.3b into Figure 8.3a? Discuss what is your expectation if you maximize β and minimize α and perform the same comparisons?

2. Discuss the same comparisons as in Exercise 8.1 for the case when α and β are of equal magnitude. Is there a relationship to the Tanimoto values for this last set of Tversky comparisons? Discuss you answer both in mathematical and visual terms.

3. The edit distance is also known as the *Levenshtein distance*. Write an algorithm that calculates the edit distance between two strings of equal length. Discuss the modifications necessary to this algorithm if the strings are not of equal length. Assume that the string lengths for your algorithm approach N, where N is a very large number. Design a modification of the initial algorithm which will be more efficient for very long sequences.

4. Consider modifications to the cost functions in Equation 8.2. Discuss the different measures that result and the impact of the cost function contributions to the utility of the resulting measures.

5. From a genetic sequence downloaded from an on-line repository, run a **BLAST** sequence similarity search and comment on the results. How similar are the two sequences being compared? As a thought experiment, comment on the similarities and additional complexities required by sequence similarity algorithms aligning protein sequences instead of genetic sequences. What percent sequence homology would you be comfortable with in moving on to building a homology model of an unknown protein structure from a known protein sequence?

6. Explore reciprocal neighbors and strongly connected components in the context of asymmetric clustering.

Chapter 9

Ambiguity

Clustering ambiguity is a little studied problem in cluster analysis, however, it has wide ramifications for results of certain clustering applications in both drug discovery and bioinformatics. Discrete features combined with common similarity coefficients create ties in proximity. Binary data and bounded count data and their associated measures typically generate a limited number of possible values, where in a given data set there are likely to be ties. Such ties in proximity are by in large the generators of a substantial amount of clustering ambiguity, either directly or indirectly. More generally, various decision steps within clustering algorithms are sometimes subject to decision ties, generating additional clustering ambiguity. Given that arbitrary decisions may be made, such algorithms are not in general stable with respect to the input order of the data. Several common algorithms and their respective decision ties are discussed in this chapter, such as, Taylor-Butina leader algorithm, several hierarchical algorithms, Jarvis-Patrick, and a common form of K-means, however it is not hard to find other algorithms that exhibit ambiguous behavior by simply inspecting their implementation for arbitrary decisions - namely, arbitrary tie breaking conditions of some form or another. As examples of discrete data, to better understand ties in proximity, we more fully explore several common similarity coefficients for binary data, such as the Euclidean, Soergel (1-Tanimoto similarity), and cosine (1-Ochiai similarity) coefficients discussed in Chapter 2.

The greater the length of a binary string in general increases the number of possible values, regardless of the measures used to generate (dis)similarities. But as the binary string length grows, it turns out that for some common measures, the number of possible values grows at a rate that may not even keep up with the quadratic nature of pair-wise similarities needed to generate a proximity matrix for clustering. Consider as a counterpoint continuous valued data, where for most clustering algorithms the results are invariant to input order, largely as a result of the fact there are no ties in proximity. One such measure is the shape Tanimoto [56] measure. Given most algorithms this continuous valued measure between shapes at significant precision generates clustering results that are almost exclusively non-ambiguous and invariant to input order. However, Taylors exclusion region algorithm has one decision tie possibility with the use of continuous data—the only such decision tie possibility among the algorithms discussed, showing, though rare, that even the use of continuous data are not immune to clustering ambiguity.

It should be noted however that clustering ambiguity is not confined to drug discovery and bioinformatics, but can be found in general, most often whenever an application clusters fixed length binary bit strings, or discrete data such as variables composed of finite counts, using the various common similarity coefficients for these types of data. Insufficient precision in either empirical data or derived data, and the possible decision peculiarities of the odd clustering algorithm like Taylor's are the other reasons that clustering ambiguity may arise in other application domains.

9.1   Discrete Valued Data Types

One common misunderstanding is that clustering ambiguity and more specifically ties in proximity result from duplicate data. For example, it is quite common in binary bit string data of any significant size that there will be duplicate fingerprints, especially if the length of the binary string is relatively small. In chemical fingerprints, the same fingerprint may represent more than one compound. This is a fact common among chemical fingerprints such as daylight fingerprints and the MDLI MACCS key fingerprints—that, not all structural nuances and differences are captured by the binary encoding, regardless of the form of encoding, even n-gram analysis of the direct encoding of the structure (e.g., SMILES stings) [57]. Depending on the fingerprint length and the compound data set there is often between 5 and 15 percent fingerprint duplicates, even with fingerprints of length 1024 bits [14]. Such duplicates will naturally cause ties in proximity, but removal of unnecessary duplicates suffices. It is the other forms of ties in proximity and decision ambiguity that we are concerned with in what follows.

9.2   Precision

Another problem relates to the precision of the data (dis)similarity coefficient values. Ties in proximity can be derived from the use of small precision as mentioned before. Typically, however, the use of the common floating point type on most modern computers eliminates the chance of ties due to precision. It can be shown, for example, that using the Tanimoto measure with fingerprints of size 1024 will not produce ties in proximity due to precision (e.g., given the 32 bit float data type in the C programming language). Providing for sufficient precision removes this as a cause of ties in proximity and clustering ambiguity. This again is not a significant issue and can in most cases be rectified pretty easily. The precision of empirical data are primarily an instrument or recording device, or data collection issue. Before clustering data in general it is simply good practice to be aware of the precision of the data and any subsequent transformation, scaling, normalization, or truncation upstream that might reduce the precision to such an extent that it has an impact on the clustering process.

9.3 Ties in Proximity

The simplest clustering ambiguity relates to the problem of a tie in proximity (see pages 76-79 in [76]) when forming a group from other smaller groups or singletons. In Figure 9.1 there are three compounds [10]. Binary fingerprints with 768 bits were generated for each compound, and measures of similarity were computed between the fingerprints. The two structures at the bottom of the triad have the same normalized 1-Euclidean and, incidently, the same Tanimoto similarity (0.919313 and 0.951923 respectively) to the structure depicted at the top of the triad. These compounds could easily be substituted for groups of compounds that are to be merged by a group proximity criterion, where there may exist a tie in the group proximity. The ambiguity arises if an algorithm merges only one of the equidistant structures or groups. Typical implementations of clustering algorithms treat ties arbitrarily (e.g., given a choice among many where only one will suffice, the first data item or group in the list is chosen), since, in general, with continuous data such ambiguity is

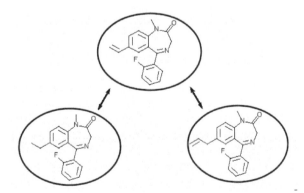

FIGURE 9.1: A common example of a tie in proximity: Given a 768 length binary string representation, a compound is equadistant, with a normalized 1-Euclidean similarity of 0.919313, as well as Tanimoto similarity of 0.951923, from two similar structures.

so rare. Given discrete features however, such as chemical fingerprints, or any binary string utilized in bioinformatics, they can be in fact be quite common [98].

The normalized Euclidean and the Tanimoto measures have been quite common as measures of (dis)similarity within the drug discovery community for a good many years, in part because they are simple and intuitive to understand, and also fast to compute. Both can be generalized to count data, and indeed there are fewer ties in proximity with count data. However, if the maximum number of counts is small and many of the counts are just one or a few per feature, these measures perform much the same way, with only slightly less ties in proximity overall from their binary counterpart.

9.4   Measure Probability and Distributions

The distribution of the measure operating on a discrete data type can impact clustering results. In Chapter 2 the distributions of measures operating on all possible pairs of binary strings of length 5 and length 6 were shown (Figure 2.5 and Figure 2.6). These distributions over the entire binary string space are somewhat misleading, in that for a given application data set, the number of binary strings is a tiny fraction of the possible number of strings, and as a sample distribution may be quite different from the entire set. In past literature, researchers used histograms or density plots to visualize these distributions, often obscuring the true discrete and often fractal nature of the distributions [17]. Such depictions may not have been detrimental to the research at hand at the time, but lent to the overall notion that these distributions behaved much like continuous valued distributions.

9.4.1   Number Theory in Measures

Motivation and theoretical basis molecular structures are often encoded in the form of binary fingerprints for a number of reasons: fast database searching and retrieval; similarity searching, clustering, and predictive modeling. They can be compared with binary similarity coefficients such as the Tanimoto similarity coefficient among many others. All of the common similarity coefficients operating on fixed length binary fingerprints have the property that the number of possible coefficient values is finite. The number of possible values depends on the coefficient and the length of the binary string. A simple example is to generate the bit strings of length 5, of which there are 32, and calculate the Tanimoto coefficient among all pairs of these strings. There are only 10 possible values that the coefficient can generate among any pair of strings. In order of their respective magnitude, these are: 0, 1/5, 1/4, 1/3, 2/5, 1/2, 3/5, 2/3, 3/4, and 1. This finite sequence is composed of all of the reduced natural number fractions, where the denominator is less than or equal to 5, and the numerator is never greater than the denominator. Such a sequence of fractions is called the Farey sequence N, where N is the largest denominator – in this example the number 5. There are however 32 strings and therefore 32 choose 2, or 496, possible unique pairs of strings. In Figure 2.5, the distribution of those values is plotted.

It is a discrete, very coarse grain distribution, and the relative frequency of values is not the least bit uniform. The situation improves somewhat as the length of the string grows large. In fact, we can calculate the approximate number of possible coefficient values based on the length of the string, given the Tanimoto coefficient. This turns out to be the sum of the Farey sequence N, where N is in this case the number of bits in the bit string, and whose asymptotic expected value is

$$\frac{3N^2}{\pi^2} + O(N \log N)). \tag{9.1}$$

For a binary string of length 1,024 bits, the exact number of Farey numbers for $N = 1024$ is 318,954 possible Tanimoto values (the expected value from Equation 9.1 turns out to be roughly 329,000 possible values). Some coefficients have far fewer. The Euclidean distance, operating on a pair of bit strings of length N, can only generate $N + 1$ possible values. The cosine coefficient is discrete, and has more possible values given N than either the Euclidean or Tanimoto coefficient, but there is no known method for calculating or approximating the number of values as N increases. In Figure 9.2, we show the possible values for the Ochiai (1-cosine), given all bit strings of length 5. There are 14 such values. A rough empirically derived estimate suggests that there are several million values, if $N = 1,024$.

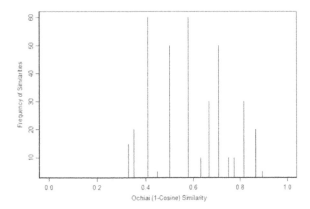

FIGURE 9.2: Distribution of all-pairs Ochiai (1-cosine) coefficient values of bit strings of length 5.

The asymmetric measure such as Tversky, in effect a parameterized Tanimoto, can produce more than double the number of unique similarity values (the exact number depends on the choice of parameter values). This increase is of little consequence in clustering ambiguity over the Tanimoto measure, since asymmetric clustering algorithms typically operate on the full (dis)similarity matrix. In addition, only certain ranges of parameters have reasonable distributions of similarity values that are particularly useful in asymmetric clustering. In the above example, all bit strings of a certain length are used to generate the total number of possible values. In applications where the bit strings are much longer, say, in the hundreds or thousands, clearly, for any data set, only a very tiny fraction of all possible bit strings of length N are used (e.g., 2N possible bit strings when N = 1,024 is astronomical compared to any reasonable data set). In addition, those bit strings typically are from a narrow and specific distribution among the population of bit strings of length N. Thus, in general though there are roughly 329 thousand possible Tanimoto coefficient values for bit strings of length 1,024, only a small fraction of those are generated by a specific data set. For example, given a data set of 100 thousand fingerprints, the total number of unique Tanimoto values could well be just 10% of the total possible. And these will not be uniformly distributed: therefore, there will be some proximity values that are far more prevalent than others. In Figures 9.3-9.5, the discrete distribution of all pairs coefficients for the Tanimoto, Ochiai and Baroni-Urbani/Buser coefficients are shown, from

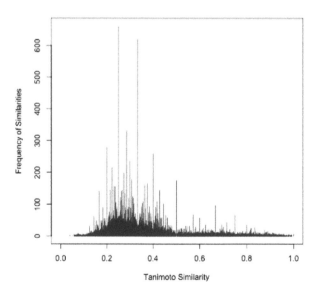

FIGURE 9.3: All-pairs Tanimoto coefficient values for 405 benzodiazepines, using Mesa 768 bit fingerprints. Of the 81,810 pairwise similarity values, given the 405 binary strings, there are only 3,678 distinct values.

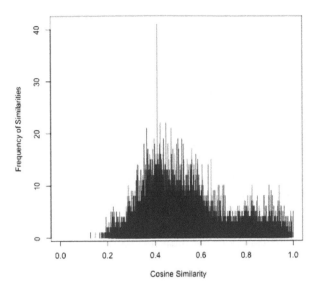

FIGURE 9.4: All-pairs Ochiai (1-cosine) coefficient values for 405 benzodi-azepines, using Mesa 768 fingerprints. Of the 81,810 pairwise similarity values, given the 405 binary strings, there are 20,332 distinct values.

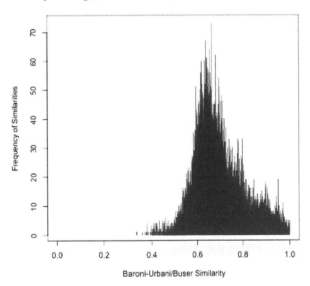

FIGURE 9.5: All-pairs Baroni-Urbani/Buser coefficient values for 405 ben-zodiazepines, using Mesa 768 bit fingerprints. There are clearly more distinct values than the Tanimoto values for this data set, but somewhat less than those of the Ochiai (the largest frequencies are nearly double that of the Ochiai values).

a data set containing 405 benzodiazepine compounds, encoded using Mesa fingerprints of length 768. Note the large frequency spikes at simple fractions such as 1/3, 2/5, 1/2, 3/5, and 3/4 in Figure 9.3 which contains the Tanimoto values. Figure 9.4 with the Ochiai values shows a somewhat more uniform distribution, and the relative sizes of the frequency spike are far less. Figure 9.5 with the Baroni-Urbani/Buser values is closer in character to the Ochiai distribution than the Tanimoto distribution in Figure 9.3.

The lesson here is that, given any reasonable length fingerprint (N), it is unlikely that all possible values are used, and the values that are used tend to be distributed towards the simple fractions. Ties in proximity are therefore the norm for common binary string coefficients, each having their own discrete size and distribution properties dependent on N. Researchers plotting these discrete coefficient distributions will find that the distributions' basic nature is little changed by the size of the data, the length of the fingerprint, or the level of compound diversity of the data set [98, 153]. Often such distributions appear in the literature as histograms or as smoothed probability density estimates, obscuring the discrete and often fractal nature of such distributions, and unfortunately giving the impression that the distributions are continuous.

Ties in proximity can quite naturally propagate to the measures used in grouping, or group merging, criteria. The clustering algorithm, Wards, for example, uses a squared error merging criterion, but, given that the values used to form the square error are from a finite source, the possible number of square error measure values is also finite, however combinatorially explosive. This can lead to ties in merging groups. A more global perspective of ties in proximity with respect to clustering binary fingerprints is thus: the use of the common similarity coefficients is typically within a space that is a rigid lattice, however remarkably fine grained or complex it may be. The space is not a continuum. Figure 9.6 shows a very simple schematic drawing in the form of a two-dimensional proximity example.

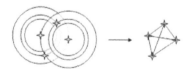

FIGURE 9.6: Lattice of discrete similarity.

On the right, the concentric circles around two of the star-like points can be thought of as possible measure values. These are not necessarily uniform in proximity or probability. Other points can only lie on the proximity circles. These points in turn have the same concentric proximities (not shown). On the left, the all-pairs proximities are drawn as line segments. The all-pairs proximities are data that a clustering algorithm operates on, and, aside from the ambiguity that arises from ties in proximity, the algorithm is independent of the rigid lattice nature of the space.

9.5 Algorithm Decision Ambiguity

It must be noted that clustering ambiguity problem is distinct from two other clustering issues mentioned above.

In the case of an agglomerative hierarchical algorithms, arbitrary decisions related to the treatment of ambiguous ties in proximity lead to the generation of a dendrogram with ambiguity resident at those levels of the dendrogram where the arbitrary decisions were made: such decisions change the underlying topology represented by the dendrogram, and in many cases changes the possible set partitions that can be derived from the dendrogram. Thus, for every arbitrary decision, a conceivably different dendrogram, and therefore topology, can be generated. For each arbitrary decision there is a separate and unique dendrogram, and simply reordering the input can generate a different dendrogram. Considering the set of possible dendrograms, the partition defined by any particular level may or may not be the same among the dendrograms. This naturally can change the results of any level selection technique across the set of dendrograms. Figure 9.7 shows schematically a single unique dendrogram (symbolically represented as a triangle) that one might generate from continuous data versus the possible multiple dendrograms that one might generate from discrete data via ambiguous decisions. A cut at any level k, represented by the horizontal line that cuts across all of the dendrograms, may produce different partitions.

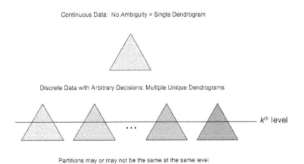

Continuous Data: No Ambiguity = Single Dendrogram

Discrete Data with Arbitrary Decisions: Multiple Unique Dendrograms

k^{th} level

Partitions may or may not be the same at the same level

FIGURE 9.7: Dendrogram (represented here as a triangle) generated by continuous data versus multiple dendrograms generated by ambiguous decisions with discrete data.

Such simple ambiguous decisions are not confined to just agglomerative hierarchical algorithms. Other common clustering algorithms are equally susceptible to ambiguous decisions. An example clustering algorithm decision may be in the form of: Take the one element or group with the minimum X value and perform operation Y. However, there may be more than one such minimum, each having a different impact on the operation Y. The possibility of arbitrary decisions in this context makes these algorithms in general unstable with respect to input order, a common source of complaint and puzzlement among researchers using these algorithms, since the algorithms are typically stable when used with continuous data. Given the same data set and features, a comparison can be made to distinguish the relative ambiguity among the clustering results of the different algorithms. Several questions arise as a result: First, is the ambiguity within a result nontrivial? And secondly, if it is nontrivial, can we still use this result within the context of the application at hand? These questions are somewhat subjective, but assume that if there are only a very few minor changes due to ambiguity in the results, then the ambiguity is trivial. In many instances however this is not the case for data sets of any significant size (e.g., 100 elements)[98]. Also, there may be a significant amount of ambiguity, but not necessarily in terms of use of clustering in the application at hand. A simple chemical diversity application can act as an example: here clustering is used to choose representatives or centrotypes from groups, and is therefore largely independent of the confounding nature of clustering ambiguity as it relates to all of the members and size of each group.

It is clear that all of the above mentioned algorithms can produce ambiguous clustering results, but the larger issue is whether the level of ambiguity can be quantified, and is such ambiguity significant from the standpoint of drug discovery applications?

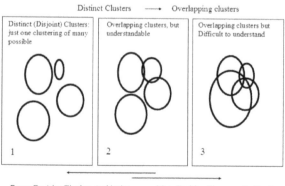

FIGURE 9.8: The three diagrams represent schematics of varying degrees of decision ties and their impact on the ambiguity of the clustering results.

In Figure 9.8, the three diagrams represent three cases where in diagram 1 the clusters are disjoint, to diagram 2, where there is only a small amount of overlapping, to diagram three, where there is a great deal of overlapping and ambiguity, and where it would be difficult to distinguish groups. Given ties in proximity and decision ties in general, a result like diagram 1 is unlikely, and the situation in diagram 2 is preferable to diagram 3. There is a need to distinguish between results that diagram 2 and diagram 3 describe. More generally, is there a quantitative measure that can be designed that identifies whether the ambiguity is significant in either relative or absolute terms.

A decision tie within the operation of a clustering algorithm presents an arbitrary change in the partition of the data set or in the construction of the topology represented by a hierarchy. The number of possible disjoint cluster-ings turns out to be the Bell number and it is exponential in the number of elements of the set to be clustered, N. One possible measure of ambiguity is to divide the log of the number of decision ties, and hence partitions, with the log of the Bell number for N. Decision ties however may not be one to one with changes in partitions. Namely, successive decision ties within an algorithm may in fact arrive at the same partition, as from a different set of decision ties. This can be seen most readily within hierarchical algorithms, where mi-nor changes in decision ties may have no impact at much higher levels of the hierarchy. If the hierarchy is partitioned via level selection, there are only n-1 possible partitions per hierarchy. Also, the number of possible topologies as defined by the hierarchies happens to be the total number of possible rooted, and leaf-labeled, binary trees with N labels; namely,

$$(2N - 3)!. \tag{9.2}$$

This number grows at a rate that exceeds the Bell number for any sizeable N (at N = 4, both sequences are 15, but at N = 5, the Bell number is 52, but the number of trees is 105, etc.). Thus there are many more trees than possible partitions, so the trees must duplicate partitions. However, for any one tree there are only a very small number of partitions that can be obtained via level selection versus the total number of partitions. It is impractical to calculate the Bell number for common data set sizes and compare them with the number of decision ties that may result in a change in the partition. Another possible ambiguity measure is to simply sum the number of decision ties, normalized to N, with the understanding that this measure is over counting with respect to the total number of possible partitions. Lastly, we can compose a measure that compares the disjoint clustering obtained by making arbitrary decisions with the number of data elements involved in overlapping clusters found by using the decision tie-based overlapping version of the clustering algorithms.

Ties in proximity can quite naturally propagate to the measures used in grouping, or group merging, criteria. The clustering algorithm, Wards, for example, uses a squared error merging criterion, but, given that the values used to form the square error are from a finite source, the possible number of square error measure values is also finite, however combinatorially explosive.

This can lead to ties in merging groups. A more global perspective of ties in proximity with respect to clustering binary fingerprints is thus: the use of the common similarity coefficients is typically within a space that is a rigid lattice, however remarkably fine grained or complex it may be. The space is not a continuum.

Clustering algorithms are based on a series of decisions that form groups. Given the discrete nature of binary fingerprint data and their coefficients, decisions are possibly faced with arbitrary choices. The number of arbitrary choices made while clustering a data set is combinatorially related to the overall ambiguity of the clustering results. This fact relates to the data input order stability of the algorithm: namely, that in a stable algorithm the output results should be invariant to the input order of the data. For example, Wards clustering is stable with respect to continuous measures and distinct data. It is not stable given binary fingerprints and the associated similarity coefficients.

9.6   Overlapping Clustering Algorithms Based on Ambiguity

9.6.1   Modified Taylor-Butina Algorithm

The first algorithm we discuss in detail is Taylor-Butina leader algorithm [132, 22], often used for cluster sampling or clustering large data sets such that can be found in compound diversity studies for compound library design. Below we list the various steps of this algorithm and identify those steps that contain the possibility of an arbitrary decision.

1. Create thresholded nearest neighbor table.

2. Find true singletons: all those data items with an empty nearest neighbor list.

3. Find the data item with the largest nearest neighbor list (thereby becoming the centrotype). There may be more than one such data item namely, there is more than one such data item with the same number of items in its respective nearest neighbor list. We will call this decision tie an exclusion region tie.

4. The chosen data item from 3 and the data items from its nearest neighbor list form a cluster, and all are excluded from consideration in the next step—the centrotype and the data items from the respective nearest neighbor list are removed from all remaining nearest neighbor lists.

5. Repeat 3 until no data items exist with a non-empty nearest neighbor list.

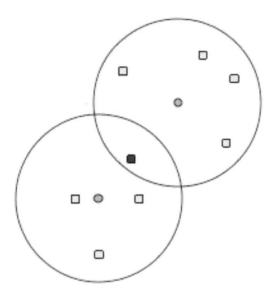

FIGURE 9.9: Exclusion region tie in proximity.

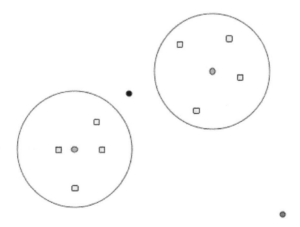

FIGURE 9.10: False singleton tie in proximity.

The exclusion region tie can cause a significant amount of ambiguity within the clustering results if the exclusion regions overlap. An overlapping exclusion region tie is shown schematically in two dimensions in Figure 9.9. The arbitrary choice of either region as the cluster to remove from the data set will almost surely impact the subsequent partitioning of the data. If this choice is based on the input order of the data items, stability of input order is lost. Since the decision is based on the largest integral value of neighbors, even if the proximity measure is continuous there is a significant chance for the occasional exclusion region tie among data items whose similarity is of a continuous nature. One of the drawbacks of this algorithm is that at completion there may be remaining data items with empty nearest neighbor tables. These we call false singletons in that they do have neighbors within the threshold, but these neighbors have been rounded up by other clusters. This problem can be addressed with the following option: 6. assign remaining false singletons to the cluster that contains their nearest neighbor within the threshold value. This modification however also introduces the possibility of ties in nearest neighbors in different exclusion regions for false singletons. In Figure 9.10, a two dimensional schemata depicts the condition of a false singleton tie between two exclusion regions, where the centrotype is denoted by a small circle in the center of the larger exclusion region circles, and the remaining data items are shown as stars. The false singleton clearly lies within the threshold proximity of members from each exclusion region and in this case equidistant in proximity. An apparent true singleton is also shown as lying outside any possible threshold region. Other simple modifications can be made to this algorithm that changes somewhat the nature of the ambiguity:

1. Use another criterion to break exclusion region ties. Such criteria may in turn have ties.

2. Use asymmetric measures which may change the number of ties.

3. Use the various decision ties in all of the above mentioned steps to create overlapping clusters rather than clusters formed by a disjoint partition of the data.

A simple and fast criterion for breaking exclusion ties is to sum the minimum proximity between the centrotype and the other data items within the exclusion region. This sum is compared with the sums obtained from the other tied exclusion regions. Though rare for exclusion regions containing any sizeable number of data items, this measure too may be tied and an arbitrary decision must be made to select the exclusion region.

9.6.2   Jarvis-Patrick

There are two common versions of Jarvis-Patrick [cite JarPat] clustering, Kmin and Pmin. In Kmin the algorithm proceeds in terms of a fixed length, k, a NN list, and a parameter, j, less than k. However, an arbitrary decision must be made if the kth neighbor is not unique due to ties in proximity. This decision impacts the clustering step of the algorithm: if two compound NN lists have j neighbors in common, those compounds are in the same cluster. If only one of the tied kth neighbors remains in NN list, there is the possibility that some possible neighbors in common are ignored, impacting what data items get clustered. In the Pmin form of the algorithm, again there are two parameters: k is the same as in Kmin; and p represents the percentage of neighbors, such that if the percentage of neighbors in common is greater than or equal to p, then the neighbors are in the same cluster. The source of the ambiguity is the same for Pmin, residing in that the kth neighbor may not be unique, but the parameter p may change amount of ambiguity. Both forms of Jarvis-Patrick with the use of ties in the kth NN can be transformed into algorithms that produce overlapping clusters.

9.6.3 *K*-Means

K-means clustering [43, 100] is a common clustering algorithm used in many fields to cluster relatively large data sets. It is more likely to be used with continuous data, but binary string data are sometimes employed. To use *K*-means with fingerprints, the 0, 1, bits are converted to continuous valued 0.0, and 1.0 respectively, and treated as separate variables. The algorithm steps are as follows:

- Choose k seed centroids from data set (e.g., quasi-randomly via van der Corput sequence)

- Find nearest neighbors to each of the centroids forming nearest neighbor group

- Recompute new centroids from nearest neighbor lists of old centroids.

- Repeat 2 until no neighbors change groups or some iterative threshold.

In step 2 centroids may compete for nearest neighbors in common due to ties in proximity. Given that the Euclidean distance is often used with this algorithm, the number of ties in proximity can be very high. In the K-Modal version of this algorithm, modal fingerprints are used to define centrotypes rather than the metric space notion of centroids.

- Choose k seed modes from the data set

- Find nearest neighbors to the modes

- Re-compute new modes

- Same as step 4 in continuous *K*-means

Modes in step three can be determined via a simple matching coefficient [121]. The modes again may compete for nearest neighbors in common due to ties in proximity

9.6.4   Hierarchical or Pyramidal Algorithms

A commonly used hierarchical algorithm used in chemoinformatics and elsewhere is Wards algorithm. A fast implementation of this algorithm uses RNN. RNN can also be employed similarly with two other common hierarchical algorithms, complete link, and group average. Decision ties that lead to clustering ambiguity reside directly within RNN. The core operation of RNN is to form a nearest neighbor (NN) chain until a RNN (each is the NN of the other) is found. Two questions arise in its implementation: What if there is more than one NN? And what if in turn there is more than one RNN? An NN chain assumes that each NN is unique. Given ties in proximity however, for any given data item there may be more than one NN. If the chain is altered arbitrarily due to ties, this situation gives rise to the second question, namely, the resulting RNN may also differ. RNNs may represent either data items or merged groups in the agglomerative process via the merging criterion in the implementation of this class of hierarchical algorithms. As represented in the dendrogram, arbitrary decisions are embedded where ever they occur within the tree. The amount of ambiguity is related to the similarity coefficient and the merging criterion. In complete link the RNNs are always the coefficient values and the ambiguity inherent in any dendrogram derived there from is directly related to ties in proximity of the coefficient values. Group average and Wards algorithms however contain derived merging criteria that are less likely to contain ambiguity, in large measure based on the complexity of the merging criterion. Hierarchical algorithms can in general be transformed into pyramidal clustering [2, 7, 8, 107] with the use of these forms of decision ties.

9.7   Glossary

decision tie: a numerical or categorical tie leading to an arbitrary decision within an algorithm

discrete: categorical, binary, or count data

tie in proximity: identical distance or similarity score

9.8   Exercises

1. Generate four example distributions of the Tversky similarity measure, given the same five-bit fingerprint used to show the similarity values of the Tanimoto measure in Figure 2.5, and the similarity values of the

Ochiai(1-Cosine) in Figure 9.2, using a range of α and β values. Describe which if any of the distributions are similar in character to Figure 2.5? Figure 9.2?

2. Comment on the advantages or disadvantages of using a Tversky measure with regards to ambiguity.

3. With discrete data explore the possibility of arbitrary decision making in either SOM or SOTA.

4. In reference [8], the author describes a method of creating a "systems of sets such that each set properly intersects at most one other set." Design an exclusion region clustering algorithm with that constraint.

Chapter 10

Validation

Cluster validity is a hard problem, especially for data sets of any significant size as the validity techniques are typically computationally expensive. In the literature, researchers will often include some form of cluster validity to help justify their results, even if the validity is some form of relative measure. Large data sets however prohibit the use of validation, so researchers have to rely on previous work with smaller sets within the application domain, or perform those studies themselves if the use of their clustering method is new to their application domain.

When presented with a specific partition as the result of a clustering method, from a quantitative standpoint *internally* validating that partition as a significant set of groups is based on two basic ideas that can be formalized mathematically. First, what is the general *homogeneity* or compactness of each individual cluster? That is to say, broadly, how self consistent is the cluster in terms of how similar all of its members are: cluster members should be very similar. There are naturally several common tests of cluster homogeneity that follow the statistical methods for determining data homogeneity based on the mean, variance, and the standard error. Second, how distinct are clusters from one another? Namely, what is the *separation* of clusters. The goal of course is that the results of a clustering—the partition into groups—will produce clusters that are well separated and each cluster's homogeneity is such that all the cluster members are very similar.

A combination of a measure of homogeneity and a measure separation, typically in the form of a ratio, can then be used to compare clustering results *externally* in a relative sense—the clustering $\mathcal{A}$ of the data has a better homogeneity/separation ratio than that of the clustering $\mathcal{B}$ of the data. Providing the clusterings are not overlapping, the ratio is independent of whether the results come from two very different clustering methods, or, just a slight change in an input parameter to the same method; or, in the case of a hierarchy, two different partitions of the hierarchy. Validation tests need, of course, to be modified for overlapping or fuzzy clustering results, where quantifying homogeneity and separation are confounded.

At least in the distinct partition case, the above notion of validation seems simple enough, but it must always be balanced against what is known (if anything) about the data before clustering. For example, it might be known that the number of groups is probably within some range, and the partition we create during clustering is far outside this range. Nevertheless, the homogeneity and separation may be very good at that clustering scale, perhaps even better than any such measure within the appropriate range of the data given the domain knowledge, but the clustering is of no use for the application. If a reference classification can be created from a small set by those experts in the domain application, this set can be clustered with a clustering method or with several methods, and the clustering or clusterings can be compared with the reference classification to determine if the type of the data might have clustering tendencies, given the data types associated with the data. Such procedures are often used to verify to the researchers that they can then go forward and cluster with larger data sets, given that the data appears to have some clustering tendency with respect to the reference classification. In the absence of such information however, validation must rely on other quantitative measures. Validation methods can measure the ratio or relationship of some measure of homogeneity versus some measure of the separation - and disregarding trivial values at or very near the extrema of either a cluster for each data object (a singleton per cluster), or just one cluster for all of the data.

It would certainly be of value to know also that the groups found differs significantly from those found by some random partition. Since the number of possible partitions is so large in general for any data set of significant size, it is only possible to sample such partitions and compare those partitions with the partition formed by the clusterings. Of course it will be important to create such a sample in a systematic way, one that draws from the almost certainly arbitrary distribution all possible partitions in terms of, for example, their ratios of homogeneity to separation. Some statistical confidence can thereby be gained via hypothesis testing. One may be able to have at least loose confidence at some interval of rejecting the null hypothesis or not rejecting the null hypothesis of the cluster being simply formed largely by chance, assuming the vast majority of the partitions are just that random, and the clustering's validation measure is quite distinct from those sampled.

10.1   Validation Measures

Validation methods fall into two categories. The first is the how well does the clustering exhibit the features of homogeneity and separation internally, irrespective of other partitions of the data. The second is external validation which considers other partitions of the data for comparison.

10.1.1   Internal Cluster Measures

For the given kth cluster, C_k, one measure of homogeneity, I_{C_k}, is simply the average of the similarity or dissimilarity to C_k's centroid or centrotype, $\bar{m}$.

If in Equation 10.1, $d(i, \bar{m})$ is a normalized measure ($[0, 1]$), then the resulting I_{C_k} is also normalized.

$$I_{C_k} = \frac{1}{n_k} \sum_{i \in C_k} d(i, \bar{m}), \qquad (10.1)$$

where i is a cluster member in C_k. Note that cluster members i and j, though distinct cluster members in C_k, may be identical in the feature space (e.g., remember that two distinct compounds may have the same fingerprint). As long as they are distinct members in the cluster, their respective (dis)similarity is considered in the sum. Another measure averages the all-pairs (dis)similarities between the members of cluster C_k, where n is the total number of cluster members in C_k.

$$I_{C_k} = \binom{n_k}{2}^{-1} \sum_{i \neq j \, \in C_k} d(i, j) \qquad (10.2)$$

Both of these measures are impacted by whether the clustering algorithm used to produce the clusters has a penchant for chaining (e.g., single link clustering) or, at the other end of the scale, compactness (e.g., an exclusion region algorithm). Which homogeneity measure is used may depend on both the type of clustering algorithm and of course the application data and the clustering goal. The measure in Equation 10.1 is just $O(n)$ to compute, whereas Equation 10.2 is $O(n^2)$, but nevertheless not particularly prohibitive. Other more complex measures can be computed, such as squaring $d()$ and taking the square root of the resulting I_{C_k} to take into account a more refined understanding of the impact of outliers in the clustering—computational resources notwithstanding. The overall homogeneity for a clustering can then be computed as the sum and average of one of the desired equations above. Given Equation 10.1, the overall homogeneity is:

$$I_C = \frac{1}{N} \sum n_k I_{C_k} \qquad (10.3)$$

10.1.2 External Cluster Measures

Separation measures are analogous to homogeneity measures. The separation measure in Equation 10.4 is simply the (dis)similarity measure between pairs of clusters' centroids.

$$E_{C_p, C_q} = d(\bar{m}_{C_p}, \bar{m}_{C_q}). \qquad (10.4)$$

Or, as in Equation 10.2, the pair-wise (dis)similarities are summed between the clusters' members

$$E_{C_p, C_q} = (n_p n_q)^{-1} \sum_{i \in C_p, j \in C_q} d(i, j). \qquad (10.5)$$

In turn we can sum and average these terms in Equation 10.3 to create an overall measure of separation of a clustering partition.

$$E_{\mathcal{C}} = \frac{1}{\sum_{p \neq q} n_p n_q} \sum_{p \neq q} n_p n_q E_{C_p, C_q} \qquad (10.6)$$

Though it may be best to regard these measures separately, we can compose a simple measure of the overall homogeneity and separation as:

$$R_{\mathcal{C}} = \frac{I_{\mathcal{C}}}{E_{\mathcal{C}}}, \qquad (10.7)$$

where, providing $d()$ in both cases is a dissimilarity, $I_{\mathcal{C}}$ tends to zero, and $E_{\mathcal{C}}$ grows, if the clusters have strong homogeneity and strong separation.

Given k clusterings, a ranking of the $R_{\mathcal{C}}$ values for each clustering will give us at least a quantitative ranking in order to determine the best clustering. If the data set is large and if k is large the computational aspects of the above equations (even as simple as 10.1 and 10.4 are) become prohibitive.

Another internal measure that can be used globally with hierarchical clustering is what is known as the *cophenetic* correlation coefficient. Given an agglomerative hierarchical clustering, the merge criterion value where the (i, j) pair of data items are first merged form $N(N - 1)/2$ values. These values form a matrix where by convention the upper triangular portion of the matrix is filled with these (i, j) values, and the rest of the matrix elements are zero. Implicitly, the matrix represents the structure of the hierarchy. How well this matrix correlates with the actual proximity matrix is a measure of the global fit of the hierarchy to the data. The cophenetic correlation can be computed simply as a form of element-wise Pearson correlation of the $N(N-1)/2$ values of each matrix (namely, the correlation between to vectors of length $N(N-1)/2$). This is also known as the Mantel statistic [101] on full matrices, where the correlation takes over the entire matrix. If the cophenetic correlation is near one, the hierarchy well represents the structure of the data. This is not to say that it is known or easily tested (Monte Carlo sampling) how close to 1 the correlation needs to be to be assured of this. The correlation tends to be larger with large data sets, and certain hierarchical algorithms may bias the correlation some, but it general it is a good first approximation as to the internal validity of a single clustering if the cophenetic correlation is above 0.8, especially for small data sets.

10.1.3 Uses of External Cluster Measures

There are three basic uses of external clustering validity, the tendency of the data to cluster, testing different clusterings, and testing clusterings against randomly generated clusterings of the same data.

10.1.3.1 Tendency to Cluster

Some work has been done to try to see if it is possible to more quickly determine whether the data has at least some significant tendency to cluster; more so than say, the faint clustering we observed in Chapter 1 concerning a uniform sampling of points. Early work concerned the use of random graph theory [13] that realistically can only be used on small data sets. This method however is still worthwhile to understand as it relates to the probability of a proximity matrix at a threshold. There are several different models of random graph generation, but the simplest [40] and the one applicable to a proximity matrix at a threshold is, $G(V, p)$, where G is the graph with the vertex set, V of cardinality n, and p is the independent probability that each possible edge in G will occur. With p set low it is quite apparent that G is unlikely to be connected, and thus be a set of components. The components represent a partition of the set V. For small V it is possible to calculate a table of the size and distribution of components given p. A curious property of random graphs in this model is that as p grows, the numbers of unconnected components are less likely, but not in a particularly uniform way with respect to the size of the components. Indeed, the most common feature is that these graphs have a single, large component (its technical terms is *Big Component*) with a distribution of much smaller components. When clustering large data sets and the resulting clusters take on sizes with this curious random graph distribution, it can give one pause. It is not necessarily the case that the data does not have a clustering tendency above and beyond a partition that would likely be formed by a random graph model, but suspicions might well be aroused.

If a clustering however does not have a common distribution of partitions formed by random graphs, then the probability is low that the clustering is happening by chance and there is good reason to assume that the clustering does show the data have a clustering tendency. For small data sets where the random graph distribution probabilities are known, one can pick a probability threshold and determine with the respective confidence whether or not the clustering partition differs significantly.

10.1.3.2 Application Dependence and Domain Knowledge

Within a given application domain, there may be enough expert knowledge to determine a test or preliminary set of classes that can be used to compare to how well a clustering method or group of methods perform. For example, given a set of compounds a group of medicinal chemists can arrive at a consensus as to a partition of the compound set into a set of classes they all agree on. The partitions that clustering methods arrive at independently can then be compared to the expert derived set of classes with a set of measures. The relative performance of the clustering methods can then inform the experts as to which method(s) might be best to use with other similar data sets [59, 148].

This procedure is quite simple. A partition of the test data of size, N, is returned from a clustering method in the form of $\mathcal{C} = \{C_1, C_2, \ldots, C_k\}$. The partition generated by the experts is $\mathcal{E} = \{E_1, E_2, \ldots, E_j\}$. Note that the number of clusters in $\mathcal{C}$ may not be the same as the number of classes in the expert partition. With $\mathcal{C}$ form an $N \times N$ matrix, $A_\mathcal{C}$, where the (i, j) matrix element represents the presence (1) or absence (0) of the i and j data members in the same cluster. Similarly, a binary matrix is generated using the expert partition, $\mathcal{E}$, $A_\mathcal{E}$. The binary matrices can then be compared much like the binary strings in Chapter 2. Only the $(N)(N-1)/2$ triangular portions of the matrices are needed as the binary matrices are each symmetric. Any of the various binary string measures can be used but typically the most common is the Tanimoto (for historical reasons, more commonly referred to as the Jaccard in this application) to capture the similarity between the clustering and the expert partition. It is then a question as to what value is sufficient to make the claim that the clustering method works well. If the clustering method is hierarchical in nature, then comparisons can be made between different levels of the hierarchy and the expert partition.

Other measures are also used in this connection, most notably Hubert's Γ statistic,

$$\Gamma = \sum_{i=1}^{N-1} \sum_{j=i+1}^{N} A_\mathcal{C}(i, j) A_\mathcal{E}(i, j), \tag{10.8}$$

and its normalized form,

$$\bar{\Gamma} = \frac{\sum_{i=1}^{N-1} \sum_{j=i+1}^{N} (A_\mathcal{C}(i, j) - \mu_{A_\mathcal{C}})(A_\mathcal{E}(i, j) - \mu_{A_\mathcal{E}})}{\sigma_{A_\mathcal{C}} \sigma_{A_\mathcal{E}} (N(N-1)/2)}, \tag{10.9}$$

where μ and σ are the mean and standard deviations of the $\binom{N}{2}$, (i, j) elements of the respective matrices, $A_\mathcal{C}$ and $A_\mathcal{E}$. Hubert's method is slower computationally, but still useful for modest sized data sets. Because of the quadratic nature of the matrices with respect to the size of the data, these methods nevertheless become unmanageable when the data size becomes large. Experts however are also limited in the size of a reference classification they are able to generate. However this form of external validity can be used to examine clusterings of the same data in a relative fashion, so their computational complexity will be an issue in this context.

10.1.3.3   Monte Carlo Analysis

Monte Carlo integration is a common tool in many fields and finds a particular utility in cluster validation, when a test for non-random structure is sought. Trying to quantify whether a set of clusters found from any given technique have some statistical significance is a non-trivial enterprise. Monte Carlo, while heavy handed in its application, is well suited for formulating large dimensional space surfaces and providing baseline comparisons of a particular space partitioning versus all possible clustering choices being chosen at random. A detailed description of the application of Monte Carlo to cluster validation will not be discussed here (see Chapter 10 Exercises), but can be easily found in other sources [76, 45].

10.2 Visualization

Computers have been a great boon to data and scientific visualization, providing almost anyone the ability to create sophisticated plots, images, three dimensional motion graphics, data animation, analysis, scientific processes, and the associated statistics and mathematics of the same. Providing visualization tools that are interpretable and especially revealing has a long history in statistics [28, 136, 137], that, in some instances, has led to the old adage that one can prove almost anything with statistics—as if statistics was simply a form of sophistry. Assuming best intentions with regards to visualization of clustering results, we nevertheless must not misinterpret or be misled by the visualizations. We are not, after all, after a fair and balanced understanding, but rather we wish to attain a best approximation of the truth as we can get *regardless*: we want as clear and revealing a view of the results as we can be obtained.

As the dimension and size of the data sets grow, the visualization problem becomes more difficult. Dendrograms provided a convenient way to visualize hierarchies, but as we have seen in Chapters 3 and 6, their interpretation becomes more and more difficult as well as the size increases. They are also confusing without careful consideration of their properties. As one simple example, we have seen that their layouts are not unique, and the novice may misconstrue the similarity of groups based on the hierarchy layout. For each clustering visualization technique there are similar associated benefits and difficulties that we find as with dendrograms. This is annoying, but part of the landscape of interpretation of results that we must navigate. Scientists have or should have a strong streak of skepticism, such that they cast a jaundiced eye at the visualization of results that suggest a confirmation of their hypotheses. Like predictive models and clustering methods, they should, if possible, use several visualization methods to help them interpret the results, e.g., dendrograms, MDS plots, heatmaps banner plots, etc.

If the visualizations reveal that the data show random noise—that is revealing! We may need go no further, or simply gain greater confirmation of this result through statistical validation. There are plenty of instances where correlation and causation are conjoined. However, there are also plenty of instances of spurious correlation that can lead to misinterpretation, mystery mongering, and other associated ills. The data may show other expected or startling results, that we can then explore with other visualization techniques and statistical studies. For all of the joy in the exploration and discovery that is science, trust the inner skeptic. Do not discount negative results. The goal of many of the figures related to clustering results throughout the text is to explore the data and possibly gain an understanding as to the validation of clustering results, whether positive or negative. Expert knowledge about the data, visualizations that help to reveal structure in the data, and validation through quantitative means as in the above sections, all go hand in hand in obtaining a more meaningful understanding of the data.

10.3   Example

A researcher is analyzing HTS results from eight screens against against their corporate collection. The researcher wants some assurance that the activities for each target have some meaning. Some compounds show high activity scores across several targets and the researcher has *a priori* knowledge that there is a relationship between targets. Can the researcher infer from the pair-wise number of duplicate compounds that the activities are meaningful? (Clustering ambiguity though possible is of little concern as the amount of data are so small.) The researcher creates a matrix of the number of duplicates per target with the total number of actives per target along the diagonal N_i, shown in Table 10.1. All N_i differ by only a small amount, but the matrix of duplicates is normalized so that the numbers of duplicates are comparable, and then the matrix is transformed into a dissimilarity matrix. The researcher clusters the data with three hierarchical algorithms, complete link, group average, and Wards. The researcher then visualizes the three hierarchies, and generates a simple heatmap with two of the hierarchies. Figure 10.1 shows the heatmap with Wards clustering on the vertical axis and Complete Link along the horizontal axis. Group average clustering dendrogram is shown on in Figure 10.2.

The cophenetic correlations between the cophenetic matrices for each of the clusterings, Wards, complete link, and group average, and the proximity matrix are 0.87, 0.87, and 0.91 respectively. This suggests that the hierarchies do a fairly good job of representing the structure of the data. The researcher however has considerable expert knowledge as to the targets and their expected correlation, and though the relationships hold up across all hierarchies, there is an anomaly, that requires further inspection. For example, target 6 is near to targets 1 and 3 which was does not correlate with their *a priori* knowledge.

The research was able to simply create clusterings, visualize the results, and generate a relative validation among several methods, as well as confirm that the hierarchies closely mirrored the structure of the data with the cophenetic correlation—and obtain valuable results, generating new information about the screen-compound process, such that the anomalous results warrant

TABLE 10.1: Matrix of Duplicate Compounds per Target

Dups Diagonal	Target 1	Target 2	Target 3	Target 4	Target 5	Target 6	Target 7	Target 8
Target 1	N1	4	76	0	0	63	5	10
Target 2	4	N2	0	15	32	0	41	8
Target 3	0	0	N3	0	0	20	1	9
Target 4	0	0	0	N4	58	0	7	5
Target 5	0	0	0	0	N5	0	6	7
Target 6	0	0	0	0	0	N6	4	7
Target 7	0	0	0	0	0	0	N7	36

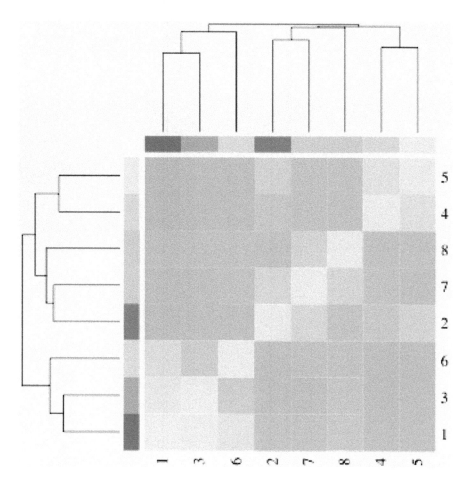

FIGURE 10.1: Complete link and Wards clustering with heatmap on duplicate proximity. Groups correspond to most of researchers expectations.

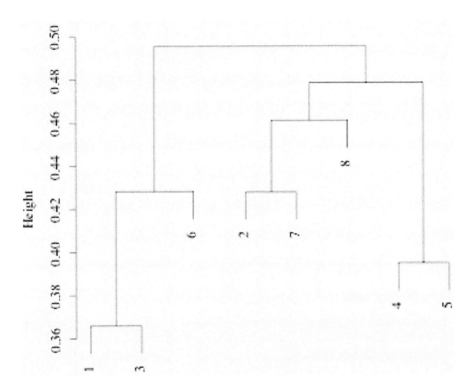

FIGURE 10.2: Group average clustering dendrogram of duplicate compound proximity. This clustering is consistent with the complete link and group average results and has the best cophenetic correlation.

further inspection. The conformation of the expected results is still speculative in some sense, but it provides some reassurance that the researchers methods are performing as expected.

10.4   Glossary

clustering tendency: whether or not a data set has internal structure at a specified scale

external clustering validity: comparing clusters to known classifications

internal clustering validity: determining the homogeneity and the separation of a clustering

Monte Carlo integration: approximating an integral with random sampling

random graphs: graphs that are generated under different probability models

---

10.5   Exercises

1. Implement Kelley's level selection [83] and apply it to small data sets clustered hierarchically [133] with several hierarchical algorithms. Compare and discuss results.

2. Outline Monte Carlo integration and in what context might it be applied to cluster validation. Search the literature for recent applications of Monte Carlo to cluster validation and discuss the merits of using Monte Carlo versus other validation techniques.

3. Hartigan suggested the following rule of thumb [63] for determining the appropriate K in K-means:

   $K + 1$ groups is a better clustering than K groups if

$$\Delta = \frac{\sum_{i=1}^{K} S_i^K}{\left(\sum_{i=1}^{K+1} S_i^{K+1}\right)(N - K - 1)} > 10 \qquad (10.10)$$

   where N is the total number of data items, and S_i^K and S_i^{K+1} are the sum of of squares for each cluster i in the Kth and $(K+1)$th clusterings respectively. Run K-means on the data sets of used in Exercise 10.1, and plot the values of Δ over a significant range for K for each data set. Compare with the level selection results in Exercise 10.1 and comment. Does $\Delta > 10$ appear to be a good rule of thumb?

Chapter 11

Large Scale and Parallel Algorithms

Large scale data sets are becoming more common in drug discovery and bioinformatics. Gene expression, genomics, metabolic pathways, chemical libraries, and molecular conformation data sets are all examples of data types that may require large scale clustering applications [50, 93]. Access to data through the creation and automation of experimental technologies has resulted in larger microarrays and High Throughput Screening assays which create simply more data to evaluate either as a pre-condition or post-process. Also, virtual combinatorial libraries of compounds and the conformational expansion of large libraries of compounds can be in the hundreds of millions even billions. Effectively grouping these data sets is a significant problem. Of course the question can be asked: is there a good scientific reason to cluster a giant data set? Typically cost benefit and efficiency play a role in the decision to use computational tools to add scientific reasoning to the decision making process for large scale datasets. For example, compound acquisition costs both in real dollars and in time create a sufficient reason to cluster large data sets and incorporate more knowledge into the decision making process. Compound diversity is another area where the application of clustering techniques has played a role, and though there are other non-clustering methods that can be used to identify diversity within a large set, there are often valid reasons to create diverse sets with clustering, as cluster membership can help in QSAR studies with diverse sets of compounds. It can certainly be suggested when gene expression screening data gets substantially larger, there will continue to be good reasons for clustering such data for exploratory data analysis and predictive modeling. The clustering of very large sets of molecular conformations is more speculative at this point, but they may help researchers organize shape data for searching and fast lookup.

Exclusion region algorithms are parallelizeable with some inter-processor communication cost, as is K-means. The caveat is that exclusion region algorithms need to have all-pairs similarities calculated as a pre-process, even if only a sparse matrix is eventually stored. Fortunately, this is embarrassingly parallel, and can be done on large clusters of processors and cores effectively. However, internal memory for storing even the resulting sparse matrix and spreading that across processor memory in a load balanced form is a nontrivial problem. If shared memory is available across all processors and cores, the memory may have to be very large accordingly. K-means and leader algorithms do not have nearly these memory constraints, as the data are effectively stored in linear and not quadratic space. Leader algorithms are sufficiently fast and space efficient to be used to cluster large data sets on a single core. Their parallelization is straightforward from a data parallelization standpoint, much like exclusion region and K-means algorithms - subsets of the data are broadcast to the cores or processors at the start of the algorithm. That data may be manipulated and changed, but very little inter-processor communication of data subsequently transpires. Certain forms of K-means are easily parallelizable, though data types and their associated measures can somewhat hobble the process, especially if they are not continuous valued with the Euclidean distance as is common in many K-means applications: there are geometric solutions such as single axis boundarizing as mentioned in Chapter 4 that can be employed with continuous valued data and the Euclidean distance that takes advantage of the finding near neighbors doesn't necessarily translate well or at all to other data types and measures.

11.1   Leader and Leader-Follower Algorithms

To start in both Leader and Leader-Follower algorithms the data are divided and broadcast to each processor in a load balanced form. In a master/slave relationship, the master sends the first randomly chosen data item to act as the first centroid to all of the slaves. Inclusion or exclusion in the cluster is determined and the results are passed back to the master. The slaves update those data items that remain and are not excluded (included in a cluster). The master maintains the clustering information and determines the next centroid or centrotype after each pass through the data. The Leader-Follower algorithm has the added twist of updating the centroid per data item, which adds considerable communication and latency. This process may be better done in a batch form, but then this becomes more and more like parallel K-means.

Algorithm 11.1 *ParallelLeader(DataSet, Threshold)*

1: **if** *Master* **then**
2:     *SubsetData(DataSet)* //Master subsets data
3:     *BroadcastSubsetData(DataSubSet, Threshood)*   //Send subsets of data to all slaves
4: **end if**
5: **if** *Slave* **then**
6:     *Receive(DataSubSet, Threshold)* //Receive data subset and threshold
7: **end if**
8: **repeat**
9:     **repeat**
10:       **if** *Master* **then**

11:     *DataElement* ← *ChooseDataCentroid(DataSet)* //Choose a
        data element from the data
12:     *Append(M, DataElement)* //Append the newest centroid to the
        list of centroids, *M*
13:     *Append(DataElement, C)* //Append the DataElement to new
        cluster
14:     *Send(DataElement)* //Send the random data element as the most
        recent centroid
15:     *Append(Receive(C'), C)* //Receive from slave subset of cluster, *C'*
16:   **end if**
17:   **if** *Slave* **then**
18:     *Receive(DataElement)* //Receive data element as new centroid
19:     *Append(FindAllNN(DataElement, DataSubSet, Threshold), C')*
        //Find all *NN* to centroid and append to *C'*, and mark all excluded
        data items in slave's data subset.
20:     *Send(C')* //Send the slave's excluded data items as cluster mem-
        bers to be appended to cluster by Master
21:   **end if**
22:   **until** *Epoch*
23:   **if** *Master* **then**
24:     *Append(G, C)* //Append most recent cluster *C* to *G*
25:   **end if**
26: **until** *SlaveDataSubsSets* == ∅
27: *Output* : (*G, M*)

Note that in line 18 of the algorithm the slave has to determine if the
data element sent as a centroid is in its data subset, and if so, mark it as
excluded. The condition *Epoch* refers to all of the slaves reporting on one
pass through their respective data. Slaves also have to report whether they
have any unexcluded data left. Naturally, the data indices are communicated,
not the data items in their entirety.

11.2 Taylor-Butina

The Taylor-Butina algorithm (Algorithm 3.5) as described in Section 5.1
can be parallelized, under the assumption that the sparse matrix is in near
neighbor table form, and that this matrix can be partitioned in to p row bands,
where p is the number of processors or cores. Let the number of rows in the
matrix be N, then a row band contains $\lceil \frac{N}{p} \rceil$ rows of the matrix. If p does not
divide N, the last row band will contain fewer rows, but this is typically little
consequence. In the event that the sparse matrix is large, either the system
has large shared memory or the processors have sufficient RAM to contain a

row band (or if the processor has multiple cores, sufficient RAM to contain as many row bands as there are cores). The algorithm is somewhat analogous to parallel sorting, where subsets of the data are processed, then merged and reprocessed. In principle, parallel Taylor-Butina has a linear speedup in the number of processors or cores. However, the algorithm works in a master-slave processor relationship, and there is some communication cost associated with marshaling sub-results from the processor slaves with a master processor, and message passing from master to slaves.

Algorithm 11.2 *ParallelTaylorButina(SparseMatrix, Threshold)*

1: *CollectSingletons(SparseMatrix, Threshold)* //Preprocessing: remove singletons.
2: **if** *Master* **then**
3:    *DistributeRowBands(SparseMatrix, Threshold)* //Slaves get row bands
4: **end if**
5: **repeat**
6:    **if** *Master* **then**
7:      *Receive(AllSlaveResults)*
8:      $C \leftarrow FindLargestSetNN(AllSlaveResults)$ //Slaves get row bands
9:      $\mathcal{G} \leftarrow Append(\mathcal{G}, C)$ //Append cluster to list of clusters
10:      *Send(C)* //Send results for Slaves to cull their row bands of C
11:    **end if**
12:    **if** *Slave* **then**
13:      *Receive(C)* //Receive results for Slaves to cull their row bands of C
14:      $C' \leftarrow FindLargestSetNN(RowBand)$ //Slaves get row bands
15:      *Send(C')* //Send row band result to Master
16:    **end if**
17: **until** *SparseMatrix == NULL*
18: **if** *Master* **then**
19:    $\mathcal{G} \leftarrow AssignFalseSingletons(\mathcal{G})$//Assign false singletons to clusters with nearest neighbors
20:    *Output : $\mathcal{G}$*
21: **end if**

11.3   *K*-Means and Variants

The *K*-Means algorithm (Algorithm 3.1) as described in Section 4.1 is relatively simply parallelizable. The core of the algorithm is to find the near neighbors of *K* centroids. This effort can be spread out among the processors again in a master-slave relationship, where the master doles out $\lceil \frac{N}{p} \rceil$ data items and the current *K* centroids, **M**, to each of the slaves. With their subsets of the data, the slaves find the nearest neighbors—namely, which data items are assigned to which centroid. The centroids then need to be recalculated. Again, in principal this algorithm should have a linear speedup in the number of processors. There of course is a latency period at each iteration while each processor finishes its task. The master calculates the new centroids once all of the assignments has been made, but it first has to assemble all of the near neighbors for each current centroid from each slave before doing so.

Algorithm 11.3 *ParallelKmeans(Data, K, Iterations)*

1: $\mathcal{G} \leftarrow KEmptyLists()$
2: $M \leftarrow ArbCenters(Data, K)$ //Find K arbitrary centers within Data
3: **if** *Master* **then**
4:    $DataRows \leftarrow PartitionData(Data)$ //Partition data into $\lceil \frac{N}{p} \rceil$ rows
5:    $Send(M, DataRows)$ //Send centroids and data rows to respective processors
6:    $i \leftarrow 1$
7: **end if**
8: **if** *Slave* **then**
9:    $Receive(DataRow)$ //Receive data row
10: **end if**
11: **repeat**
12:    **if** *Master* **then**
13:       $\mathcal{G}' \leftarrow \mathcal{G}$
14:       $Receive(C)$ //Receive near neighbor assignments of data from Slaves
15:       $\mathcal{G} \leftarrow AssembleNN()$ //Assemble NN groups from all C to calculate new centroids
16:       $M \leftarrow FindCentroids(\mathcal{G})$ //Calculate new centroids
17:       $Send(M)$ //Send new centroids to Slaves
18:       $i \leftarrow i + 1$
19:    **end if**
20:    **if** *Slave* **then**
21:       $Receive(M)$ //Receive new centroids
22:       $C \leftarrow FindNN(DataRow, M)$ //Find the nearest neighbors to M
23:       $Send(C)$ //Send near neighbor assignments of data to Master
24:    **end if**
25: **until** $i == Iterations$ OR $\mathcal{G} == \mathcal{G}'$
26: *Output:* $\mathcal{G}, M$

11.3.1 Divisive *K*-Means

Divisive *K*-means is nearly as simple to parallelize, provided that there is either shared memory or there is sufficient memory per core such that the data can be stored in full. Unlike those clustering methods that need a full or sparse proximity matrix, only the data need to be stored and thus are linear and not quadratic space, so even cores with only a modicum of available RAM can store millions of data items. Individual *K*-means jobs can be sent to cores in the divisive sequence, where the cores accept jobs in a queue formed by the master and report results to a master. The simple stopping criterion is of course letting the process divide groups until the leaves of the hierarchy are all single data items. This is somewhat self-defeating as a considerable amount of computation may be done to little purpose. Local criterion can be applied to each group, or more global measures of the hierarchy can be applied at each step.

Algorithm 11.4 *DivisiveHierarchical−K−means(Data)*

1: $N \leftarrow Size(Data)$
2: $DistributeData(Data)$ //Distribute data to all processors if not shared memory
3: $QueueGroup(Data, Q)$ //Queue the data as the first group.
4: **repeat**
5:    **if** *Master* **then**
6:       $Send(\mathcal{G}, M, Q)$ //Send centroids, M, and group $\mathcal{G}$ from Q to be divided
7:       $Receive(\mathcal{G}_1, \mathcal{G}_2)$ //Receive pair of groups from a slave
8:       $QueueGroup(\mathcal{G}_1, Q)$ //Queue new group
9:       $QueueGroup(\mathcal{G}_2, Q)$ //Queue new group
10:      $Stop \leftarrow ComputeStop()$ //Compute stopping criterion
11:    **end if**
12:    **if** *Slave* **then**
13:       $Receive(\mathcal{G})$ //Receive group, $\mathcal{G}$, from master
14:       $(\mathcal{G}_1, \mathcal{G}_2) \leftarrow ComputeKmeans(\mathcal{G})$
15:       $Send(\mathcal{G}_1, \mathcal{G}_2)$ //Receive pair of groups from a slave
16:    **end if**
17: **until** $Stop == True$

11.4   Examples

11.4.1   Compound Diversity

With the adoption of combinatorial chemistry in the late 1990s, pharmaceutical companies found themselves trying to understand to what extent the synthesis of large combinatorial libraries added structural diversity to their proprietary collections [102]. Even to this day, the ideas of assessing and evaluating collections of compounds for their inherent diversity (either in structural space or property space) are ongoing [53]. Combinatorial library design itself is an exercise in compound diversity, in that the cost benefit of a synthesis of the enumeration of a scaffold with a collection of substituents is typically maximized by emphasizing structural diversity, limiting duplications or close

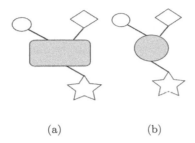

(a) (b)

FIGURE 11.1: Figure 11.1a and Figure 11.1b are a schematic representation for two combinatorial libraries with identical substituents (star, circle and diamond) but with different scaffolds (rounded blue rectangle and an oval).

similarities and maximizing the chemical space coverage with the smallest number of library members.

In Figure 11.1a and Figure 11.1b, pictorial representations of a single member of two different combinatorial libraries are displayed. The "scaffolds" are different, in Figure 11.1a the scaffold is represented by a rounded rectangle and in Figure 11.1b the scaffold is represented by a shaded oval. These "cores" define a combinatorial library series. Libraries are enumerated by adding a variety of substituents in specific positions as appendages to the scaffolds. In Figure 11.1 both of these library members have three substituents, the star, square and circle. In a typical library design the position and characteristic of the substituents once enumerated, create the potentially large combinatoral library. For example, if the three positions in Figure 11.1a were each replaced by fifty unique appendages, the library member in Figure 11.1a would come from an enumerated library of size 250,000.

Even with identical substituents, the two libraries would still create diversity to a larger collection due to the fact that the scaffolds are unique and the substituent attachments onto the unque scaffolds create potentially even greater structural diversity between the two combinatorial libraries. Diversity based on chemical series would dictate a desire to include different series scaffolds which can provide both chemical differences which reflect different drug target interaction possibilities both in terms of the structural features as well as the poses or diversity in conformational space.

At first glance, one can easily see how to increase the diversity of a combinatorial library itself, by including a variety of substituents, increasing the number of substituents and the number of links to the scaffold itself. Robotics has taken over much of the combinatorial library synthetic preparation [62]. If the chemical reactions to create the library are known, libraries of very large sizes are routine. Cost typically enters into a decision on the synthesis of such large libraries. Just because numerous compounds can be made, is there a benefit to having a large set of compounds with the same scaffold? Typically the answer to this depends on what contribution the screening of such a library will make to the drug discovery process. If combinatorial libraries are being screened for lead discovery, then typically many smaller libraries with diverse scaffolds enables larger coverage of chemical space in the quest for one series that preferentially binds to a target. If the combinatorial libraries are being utilized in the lead optimization part of the drug discovery process, then enumerating large numbers of compounds with a scaffold that has an arrangement of substituents that bind to a target, then diversifying around the scaffold with very similar structures provides valuable information for SAR and optimization of drug potency.

11.4.2   Compound Acquisition

Proprietary compound databases are the bread and butter of a pharmaceutical company's drug discovery process. Augmenting, supplementing and expanding a company's proprietary compound collection is an ongoing process for all companies that screen compounds against proprietary biological receptors.

Currently the number of available chemicals for purchase or for screening by a drug discovery company is of the order of millions. Clustering large scale available vendor databases and performing similarity comparisons to in-house databases to determine the chemical advantages of purchasing one vendor's databases over another is an area where clustering large scale data sets has an application; namely, finding a diverse sampling of compounds from the database, sufficiently unlike the in-house database, that form a subset to purchase and thereby augment the propriety collection.

However, companies with experience in screening databases, typically adhere to certain standards in terms of what types of compounds they may wish to purchase to augment their databases. Usually, in a compound assessment, a company has a better idea of the types of structures they *do not* want, versus an understanding necessarily as to the compounds that they *do* want. Much of this is due to the fact that a good deal of research has been done on determining which structural features lend themselves to covalent protein interactions, promiscuity, and other unwanted complications from screening compounds. As a result, companies will first filter any possible compounds under consideration based on either industry standard practices or create their own proprietary filters and apply these first to any vendor databases that they might purchase.

Standard industrial filters for screening compounds depend on the kinds of compounds screened: there are standard compound databases which may or may not contain precursors, fragment databases which tend to be smaller in molecular weight, combinatorial libraries, etc. and somewhat differing filtering criteria may apply to each. For first pass lead discovery, common filters include some or all of the Lipinski rule of five criteria; that is, leads from the screening process should not be too large, as lead optimization tends to increase molecular weight, they should not be too greasy or insoluble, and they should not have the potential to covalently bind to the protein. The Lipinski's rule of five refers to the fact that each rule is a multiple of five [96]:

- molecular weight of less than 500

- *clogp* less than 5

- 5 hydrogen bond donors or less

- 10 hydrogen bond acceptors or less

In general the rules have been found far too restrictive, such that they are often modified substantially to suit acompany's databases and applications. Typical filters include removing features common in receptor covalent binding, promiscuity or **ADMET** [108, 114]. Also, considerations on synthetic feasibility or lead optimization as well may be included in the initial data processing.

Once a set of company acceptable filters are in place, all vendor databases that are under consideration are filtered and then can be analyzed either through structural clustering or chemical structure similarity comparisons. Companies are looking for compounds which are somehow different from those they already own, trying to diversify their internal proprietary databases to increase their structural space that is covered in their screens. The hope is that this diversity enhancement will lead not just to viable leads, but also those leads will be optimized for potency and patentability [89].

Vendor databases range in size, from tens of thousands to millions of available compounds. Large scale clustering of 2D structural fingerprints of the compounds of a potential database under consideration for purchase enables a structural assessment. Even the 3D shape diversity can be understood via clustering multiple low energy conformations of some of the smaller data sets. Large computing resources makes this more feasible for the larger data sets.

Filtered vendor databases are available online (e.g., **ZINC** [74]). Structural clustering discovers representative compounds that span the compound offering from a given vendor.

Representatives are the cluster centroids and the singletons at a given clustering threshold, in the case of Taylor's algorithm. The threshold cutoff generates clusters based on the similarity of a set of structures to other structures based on the threshold. Threshold selection is data and fingerprint dependent. The kinds of representation that is utilized for compound/compound comparisons determines a meaningful threshold value. (See example from clustering above at a threshold cutoff of 0.75?) Much work has been described that addresses a so called good cutoff value for different fingerprint representations. A given company may have their own fingerprint or their own opinion as to what threshold generates groups of the compounds which "look" alike or are structurally similar by their chemists assessment.

Discovering the structural space represented by a given vendor database is the first step in determining the degree to which a database may be of value to a drug discovery company. The next step is to determine the distance these representatives are from a drug discovery company's compound collection. A similarity search can be performed which assesses the similarity between the representatives and an in-house database. Again a threshold cutoff is necessary to determine how close a given purchasable compound is to any compound already in the collection. The similarity assessment will determine the representatives which will add the most diversity to an in-house collection, since they are at least a threshold's difference away from any compound currently owned.

Another aspect to compound selection are lead optimization considerations. SAR (structure activity relationships) neighbors of a "lead" found from a screen are important in determining the viability of a given hit as well as for synthesis design for lead optimization. For this reason, some companies will add an additional selection criteria in that they only select compounds for purchase which also have a certain number of SAR neighbors [20]. Very close neighbors (typically 5 to 10) are chosen from the representative's respective cluster.

Large scale clustering in 2D as described above enables structural diversity enhancements of in-house databases or of screening databases for companies that outsource their screening. With the advent of 3D descriptors and the availability of large scale computing resources, companies can also make decisions on the diversity enhancement of purchase compounds in terms of how they will add to the 3D diversity of a compound collection.

Shape is one such descriptor which has been utilized in various ways in computer assisted drug design. Understanding how a compound potentially binds to a given receptor is where 3D structure and interaction is critical to binding and also lead optimization.

3D descriptors which can take on a binary form can be clustered in large scale just as straightforwardly as 2D chemical structure fingerprints. One example is the clustering of shapes based on a shape fingerprint. In 3D one of the main confounding issues in terms of computational complexity is the conformer issue. Any given structure can be represented by many conformers. Deciding how many conformers adequately represents a given 2D structure is beyond the scope of this discussion. But if a company has a 3D representation that describes, in their opinion, an adequate encapsulation of the 2D structures in 3D structural space, then one can cluster these structures to look for similar compounds in 3D. One can perform the same diversity assessment as outlined above. First, analyzing a vendor database in 3D, for example generating 3D shape fingerprints, clustering to discover shape similar groupings and then taking the representatives and searching to see if any groups differ under some criterion or criteria in shape from those already found in the in-house database. In the case of shape, companies may decide to keep compounds which are similar, or keep only compounds which have different shapes than what they already own. The thresholding makes the assessment of these decisions transparent.

11.5 Glossary

in-house collection: A pharmaceutical company's internal proprietary collection of compounds

Lipinski Rule of Five: a set of four rules for pre-filtering screening compounds, each with a threshold of the number 5.

QSAR: Quantitative Structure Activity Relationships - that biological activity is quantitatively related to chemical structural properties

SAR: Structure Activity Relationships- related structures with their associated activities

scaffold: core chemical structure which is the foundation of a combinatorial library

shape: A descriptor in 3D of a chemical conformer

vendor: Chemical compound provider

11.6 Exercises

1. Consider two databases A and B each containing over 1 million compounds. You are asked to determine which database will add the most diversity to your corporate collection C which contains 6 million compounds. Discuss what method or methods you would apply to do the analysis and what criteria you will use to determine which database is of more value to your company assuming that you were required to purchase A or B in its entirety. Describe how or if you would do the analysis any differently if you were to purchase just a subset of A or B.

2. In Taylor-Butina, the near neighbor table, even though just a sparse portion of the full proximity matrix, is space inefficient in the symmetric clustering version of the algorithm, as it stores both (i, j) and (j, i) elements. What computational impact would be required if the near neighbor table was stored as a single array with offsets without the duplicate information: 1)in the sequential algorithm; 2) in the parallel algorithm.

3. Divisive K-means purportedly returns clusters whose sizes are are more uniform than other algorithms for large scale clustering. What would be the value of such results for the applications mentioned in this chapter? What might be the problem with such results given any application?

Chapter 12

Appendices

The following appendices are meant in large part as a review of mathematical topics that the reader should have at least a passing understanding if not a deep grounding in. They are also slanted to the usage of those topics in cluster analysis. Some of the most basic elements are assumed.

12.1   Primer on Matrix Algebra

A great deal of the mathematics of statistics is in the form of linear algebra, where the manipulation of matrices and vectors is commonplace. Statistical learning theory and in particular regression theory contain many formulas and derivations in the form of vectors and matrices. Cluster analysis is often concerned with pattern and (dis)similarity matrices, their properties, relationships, and transformations. An additional important concern is the how vector and matrix data are represented in a computer and how machine operations on these data impacts results, given finite precision. This later concern is the domain of numerical analysis, which is replete with numerical algorithms for operating on vectors and matrices and arrive at results of a given precision and error.

By convention an n-dimensional vector is displayed vertically:

$$\mathbf{x} = \begin{pmatrix} x_1 \\ x_2 \\ \vdots \\ x_n \end{pmatrix}$$

Its transpose is laid out horizontally:

$$\mathbf{x^T} = (x_1, x_2, \ldots, x_n) \tag{12.1}$$

This convention follows vector–vector and vector–matrix operation conventions. The *inner product* or *dot product* of two vectors of equal length, the sum of the corresponding pair-wise entries of each vector, thereby equals a scalar value.

$$\mathbf{x}^T\mathbf{y} = \sum_{i=1}^{n} x_i y_i = a. \tag{12.2}$$

The length of a vector, $\mathbf{x}$ is

$$||\mathbf{x}|| = \sqrt{\mathbf{x}^T\mathbf{x}}, \tag{12.3}$$

and is referred to as the *norm* or the *Euclidean norm* of a vector, as it is equivalent to the Euclidean distance to the origin, if we consider the vector, $\mathbf{x}$, as the point x in Euclidean space.

The outer product of two vectors generates a matrix in the dimensions of the vectors (commonly, their lengths are the same, but need not be):

$$\mathbf{x}\mathbf{y}^T = \begin{pmatrix} x_1y_1 & x_1y_2 & \cdots & x_1y_n \\ x_2y_1 & x_2y_2 & \cdots & x_2y_n \\ \vdots & \vdots & \ddots & \vdots \\ x_my_1 & x_my_2 & \cdots & x_my_n \end{pmatrix}$$

Matrix addition on matrices of the same dimension simply adds the matrices element-wise.

$$\mathbf{A} + \mathbf{B} = \mathbf{C}, \tag{12.4}$$

such that,

$$A_{i,j} + B_{i,j} = C_{i,j} \tag{12.5}$$

Simple matrix multiplication is performed using the inner product of the rows (horizontal vectors) of the matrix on the left with the columns (vertical vectors) of the matrix on the right. The dimensions of those matrices are therefore:

$$\mathbf{A_{m,n}}\mathbf{B_{n,m}} = \mathbf{C_{m,m}}, \tag{12.6}$$

Matrices without dimensions are assumed to be square.

$$\mathbf{AB} = \mathbf{C}, \tag{12.7}$$

The element by element product of a matrix is the Hadamard product, $\mathbf{A} \otimes \mathbf{B}$ and it is distinct naturally from the matrix multiplication, $\mathbf{AB}$. The square matrix whose diagonal is all ones, and other entries are zero otherwise, is known as the *identity matrix*.

$$\mathbf{I} = \begin{pmatrix} 1 & 0 & \cdots & 0 \\ 0 & 1 & \cdots & 0 \\ \vdots & \vdots & \ddots & \vdots \\ 0 & 0 & \cdots & 1 \end{pmatrix}$$

Quite naturally,

$$\mathbf{AI} = \mathbf{A}. \tag{12.8}$$

The transpose of a matrix, $\mathbf{A}^T$, swaps the (i,j)th element for the (j,i)th element. The determinant of a matrix is given the notation, $|\mathbf{A}|$, and it represents an operation that returns a scalar value. If all the vectors of a matrix are linearly independent (their pair-wise inner product is zero), the determinant is non-zero. The inverse of a matrix, akin to the reciprocal of a real number, is denoted by $\mathbf{A}^{-1}$, providing the matrix is *invertible*: that is, the determinant of the matrix is not zero—analogous to the fact that there is no reciprocal of zero in simple arithmetic.

The concept of eigenvalues and eigenvectors are very important and often fundamental to a considerable amount of statistical learning theory. It is used in single value decomposition, principal component analysis, spectral clustering, and multi-dimensional scaling among other applications. It would be nice therefore to give a more extended presentation on this topic, thereby giving an intuitive feel for the subject, rather than just a set of definitions, equations, and identities.

Starting with the mathematics, for an $n \times n$ matrix $\mathbf{A}$, the linear equation

$$\mathbf{A}\mathbf{x} = \lambda\mathbf{x}, \tag{12.9}$$

can be solved in n ways. Namely, there are n possible vectors that $\mathbf{x}$ can take on with a corresponding n λs (that may not be necessarily unique). Namely,

$$\mathbf{A}\mathbf{x_i} = \lambda_i\mathbf{x_i}, \tag{12.10}$$

where $i = 1, 2, \ldots, n$.

To solve for x_i and λ_i, Equation 12.9 can be transformed to

$$(\mathbf{A} - \lambda\mathbf{I})\mathbf{x} = \mathbf{0}, \tag{12.11}$$

where $\mathbf{0}$ is a vector of n zeros. Now if we take the determinant of the expression $\mathbf{A} - \lambda\mathbf{I}$, where $\mathbf{A}$ is an $n \times n$ square matrix, we get the characteristic equation,

$$|\mathbf{A} - \lambda\mathbf{I}| = \lambda^n + a_1\lambda^{n-1} + \cdots + a_{n-1}\lambda + a_n = 0. \tag{12.12}$$

This equation can be solved for n, not necessarily unique roots, λ_i. With each λ_i we can solve the set of linear equations that would occur by substitution in 12.11 and derive the corresponding eigenvector, $\mathbf{x}$.

To obtain an intuition as to the mathematics we can use the application of finding the principal axis of an 3-dimensional data set, with the understanding that the application can be generalized to N-dimensions. Figure 12.1 contains a cloud of points. The points were generated via the Hammersley sequence in 3-dimensions in a cube, and a scalene ellipsoid subset of those points were selected by rejecting all the points outside of the ellipsoid. Here we have mean centered the data about the origin, and then generated the eigenvectors and eigenvalues (either via singular value decomposition on the mean centered data or via principal components analysis, using the covariance matrix of the mean centered data), and rotated the points to their orthogonal eigenvector

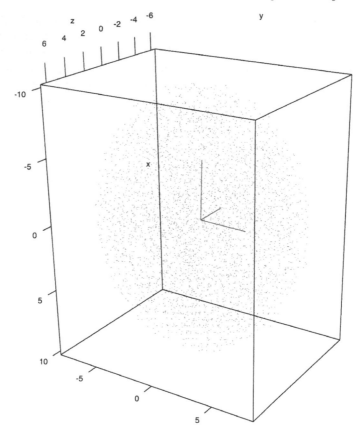

FIGURE 12.1: A 3-dimensional cloud of quasi-randomly generated points inside a scalene ellipsoid with principal axes of those cloud of points. The scale is the same as that of the molecule in Figure 12.2 The orthogonal eigenvectors are plotted along the x, y, and z axes.

basis such that they line up in descending eigenvalue order along the x, y, and z axes. The lines in the center of the points are those orthogonal eigenvectors, and their length is determined by their eigenvalues and related directly to the variance of the point set. In this instance the scalene ellipsoid has a different radius for each axis—a bit like sitting on a rugby or American football laid lengthways.

This is in a sense an ideal case. Now, say, we take a subset of those points that are not at all like the simple scalene ellipsoid. In Figure 12.2 we have a small molecule placed inside the same box, with the scale in Angstroms. Its shape is defined by a set of overlapping atom spheres. It too is mean centered by placing the centroid of the atom centers at the origin. If we then overlay the points in the ellipsoid from Figure 12.1, a subset of points will lie

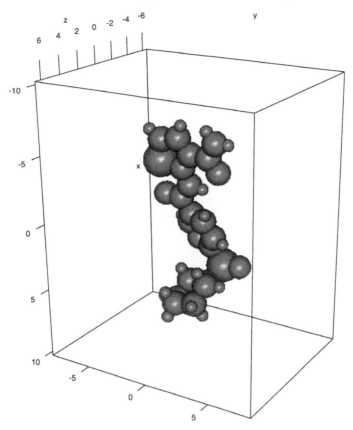

FIGURE 12.2: A 3D depiction of a small molecule with the scale in Angstroms.

within the atom spheres. This subset of points is clearly quite different. Again we can calculate the eigenvectors and eigenvalues of this subset. If we plot now those eigenvectors at the origin inside the ellipsoid as is done in Figure 12.3, we can see again that they are orthogonal and have somewhat different magnitude, as the set of points that define the molecule are quite different, but their extent which defines the variance of those points fits within as it were a different, smaller ellipsoid with different radii. Note that some of the axes point in a different direction, and that is due to the fact that the sign of the eigenvector is arbitrary. In Figure 12.4 one of four possible orientations of the 3D conformation is shown inside the scalene ellipsoid sampling region.

This method is indeed used to calculate the volume of a complicated shape via quasi-Monte Carlo integration [45, 106], wherein the ratio of the number of subset points within the shape over the total number of points, times the total known volume that encompasses all of the points is an approximation

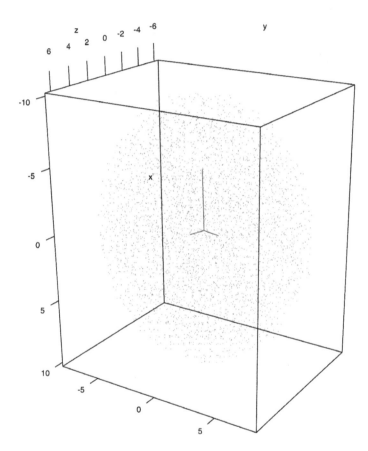

FIGURE 12.3: A 3-dimensional cloud of quasi-randomly generated points inside a scalene ellipsoid with a 3D depiction of a small molecule with principal axes of those cloud of points within the sphere, shown given their eigenvalue determined lengths. The scale is in Angstroms.

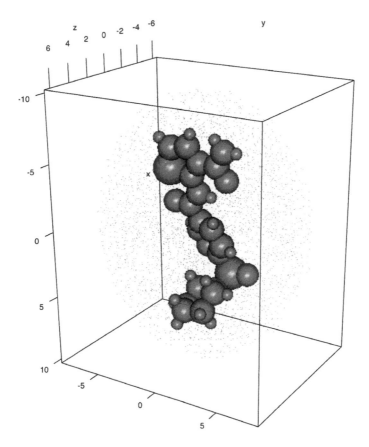

FIGURE 12.4: A 3D depiction of a small molecule and quasi-randomly generated points inside a scalene ellipsoid sampling region.

of the volume of the shape. It can also be used to align shapes (provided the sign ambiguity of the eigenvectors is removed [19]) as can often be found in the imaging literature and it can be used to compare the similarity of shapes, leading to a similarity measure of molecular shape, that in turn can be used in cluster analysis of molecular shapes.

Calculating eigenvalues and eigenvectors efficiently and accurately from the standpoint of numerical analysis is an important consideration. In MDS for example, we may only need to compute the first two eigenvalues and eigenvectors of a proximity matrix, and not all n pairs that one would calculate using singular value decomposition. There is in fact a method, known as the *Power method*, that can be used for generating the first few ($m \ll n$) eigenvalues and eigenvectors that is much more efficient than SVD.

12.2 Primer on Probability Theory

Probability theory concerns the quantitative analysis of random events and the probability distribution of random events. There are two basic and parallel notions of random variables, namely, *discrete* and *continuous*. In the discrete case, there are n possible events, each with a non-negative probability, p, of occurring, such that,

$$\sum_{i=1}^{n} p_i = 1. \tag{12.13}$$

The *probability mass function* is expressed by the discrete random variable, x, taking on the any one of the events, in the finite set of events, X, such that

$$\sum_{x \in X} P(x) = 1. \tag{12.14}$$

If x takes on continuous values, the probability can only be determined over an interval. This translates into what is known as the *probability density function*

$$\int_{-\infty}^{\infty} p(x)dx = 1. \tag{12.15}$$

Here x is assumed to take on values over the interval $(-\infty, \infty)$, and the probabilities are non-negative.

The *expectation* of the distribution of a random variable is defined as

$$\mathbf{E}[\mathbf{x}] = \sum_{x \in X} x P(x) \tag{12.16}$$

This is commonly referred to as the *mean* or *average*, and often denoted as μ, but we speak of the expectation as an operator with specific properties, and rather than just a set of values from a sample, the expectation can operate on a function. Similarly the *variance* of the distribution is defined by

$$\mathbf{Var}[\mathbf{x}] = \mathbf{E}[(\mathbf{x} - \mu)^2], \tag{12.17}$$

and denoted as σ^2. The variance gives a measure of spread of the distribution. The standard deviation is simply the square root of the variance. With the standard deviation we can measure off the distribution in terms of distance from the mean, as so many standard deviations from the mean.

The concept of a random variable and probability can be generalized to the joint probability of two or more random variables. The covariance of two random variables expressed in vector form, $\mathbf{x}$ and $\mathbf{y}$, gives the covariance matrix,

$$\mathbf{\Sigma} = \mathbf{E}[(\mathbf{x} - \mu)(\mathbf{x} - \mu)^{\mathbf{T}}]. \tag{12.18}$$

This latter expression is exceptionally useful in a great many data applications.

12.3   Primer on Number Theory

Number theory was, until the mid-twentieth century, considered an area of pure mathematics unsullied by derivative applications. Today it is widely applied in areas of cryptography, computer science, and statistical modeling. It is remarkable how a mathematics that begins with the natural numbers and simple arithmetic proceeds to exceptionally sophisticated mathematics, bridging combinatorics, topology, and analysis among other mathematical fields; and it includes some of the most difficult and celebrated unsolved problems in mathematics today, such as the enormously deep Riemann's hypothesis, and the unbelievably simple to state, Goldbach conjecture (the sum of any even positive integer greater than 3, is the sum of two prime numbers). This primer is mainly concerned with with an introduction to elementary and analytical number theory sufficient to provide a grounding in its use within the confines of cluster analysis, which are few and rather specific. One such simple concept, that of the Farey sequence, is related to Riemann's hypothesis, independent of its application in cluster analysis. At the end of this section, this connection shown for the general bemusement of the reader.

The natural numbers and simple arithmetic have generally recognized facts: there are odd and even numbers, some are divisible, some are not, there are prime numbers, perfect numbers, square free numbers, twin primes (3 and 5, 5 and 7, 11 and 13, 17 and 19...), etc. Number theory begins with these simple facts and proceeds to theorems about these and related facts and assorted arithmetic axioms and extensions of the concepts of number and arithmetic. Simple questions arise (e.g., are there an infinite number of twin primes?) that have profoundly deep mathematical solutions, or remain open - leading to a great deal of additional mathematics and more questions, simple and otherwise.

A prominent feature among many of these facts, theorems, and questions (or stated in the positive, *conjectures*, e.g., there *are* infinitely many twin primes) are the prime numbers and their distribution among the natural numbers. A few theorems and definitions include:

- Theorem: There are infinitely many prime numbers.

- Theorem: Any natural number can be factored into a unique set of prime numbers.

- Theorem: The distribution of primes in the natural numbers is bounded.

- Definition: If the prime factorization of any two distinct natural numbers contain no prime number in common, those two natural numbers are considered *relatively prime* or *coprime*. Namely, there is no common divisor between the two numbers, so they are relatively prime to one another.

- Definition: The numerator and denominator of a fully reduced common fraction (fractions between 0 and 1) are coprime to one another. The sequence of such pairs is known as the Farey sequence.

The above definitions and theorems concern the frequency and enumeration of primes and relative primes. Primes appear to occur in a strange and apparently random frequency with increasing number of natural numbers up to some large N. Mathematicians began to ask what was the nature of this frequency and could it be quantified or bounded in some way. The first real exploration of the frequency of primes is due to Gauss. He realized that there was a rough relationship in terms of the natural logarithm, called the integral logarithm, $li(x)$:

$$li(x) = \int_0^x \frac{dt}{\ln t}, \tag{12.19}$$

where x is a positive real number and $x \neq 1$. He showed that the number of primes less than or equal to large N is roughly $\frac{N}{\ln N}$. It follows also that the probability of any positive integer between 1 and N being prime is roughly to $\frac{1}{\ln N}$. This is known as the *prime number theorem* and it is an instance of analytical number theory.

A simple notion that one learns in grammar school is how to reduce a proper fraction (rational numbers between 0 and 1) to a form where they can be reduced no further. As stated above such pairs of numbers are relatively prime. A French mathematician in the 19th century described a method for generating all proper fractions $\frac{p}{q}$, where $p < q$, $q \leq n$, and p and q are coprime for positive integers, p, q, and n. This is a sequence of rational numbers called the Farey sequence order n. For large n, the expected value of $\mathcal{F}_n$ is

$$\frac{3n^2}{\pi^2} + O(n \log n)). \tag{12.20}$$

This remarkable result relates π to reduced proper fractions. In fact, it can be easily shown from this result that the probability of any two integers being coprime is

$$\frac{6}{\pi^2} \approx 0.608 \approx \frac{3}{5}. \tag{12.21}$$

Thus, roughly three fifths of the set of proper fractions $\frac{p}{q}$, where $p < q$, $q \leq n$, for large n are fully reduced fractions, and the remainder reduce to one of these fully reduced fractions. (the probability that any three integers are relatively prime is given by Apery's constant - approximately 0.832.) Given rational similarity measures, it is important to keep this fact in mind, as it may have a bearing on the form of the distribution of the similarity values, where low order fractions (e.g., $\frac{1}{2}$, $\frac{1}{3}$, $\frac{2}{3}$, $\frac{1}{4}$, etc.) are more prominent in the distribution than would be our natural expectation.

These distributions have fractal properties (ruler pattern plot) as do other prime number distributions (Ulam's square).

Other curious relationships to $\frac{6}{\pi^2}$ are:

$$\frac{6}{\pi^2} = \sum_{n=1}^{\infty} (n^{-2})^{-1} \tag{12.22}$$

$$= P((m, n) \; are \; coprime | m, n \in \mathbb{N}) \tag{12.23}$$

$$= P(n \; is \; square \; free | n \in \mathbb{N}) \tag{12.24}$$

$$= (\prod_{p=2}^{\infty} (1 - p^{-2})^{-1})^{-1} \tag{12.25}$$

$$= \sum_{n=1}^{\infty} u(n) n^{-2} \tag{12.26}$$

Equation 12.22 is Riemann's zeta function, $\zeta(2)$. A *square free number* has no power greater than 1 in its prime factorization (e.g., $2 \times 3 \times 5 = 30$ is square free; $2 \times 3^2 = 18$ is not). The expression in 12.25, the p refers to prime numbers. In Equation 12.8, $u(n)$ is the mobius function: $u(n) = 0$, if n has a square factor, else 1 if n has an even number of prime factors, and -1 otherwise.

As simple to state as the Farey sequence is, it can be used to state the Riemann hypothesis. Let the total number of fractions in the Farey sequence of order n equal K (namely, it can be determined exactly, but it is approximately on average related to the expression 12.23). For each Farey number in sequence (they are rank ordered), the ith Farey number, take the absolute difference between it and $\frac{i}{K}$. Then let the ith absolute difference be δ_i. We can the state the Riemann's hypothesis as:

$$\sum_{n=1}^{\infty} \delta_i = o(n^{1/2+\epsilon}),$$ (12.27)

where epsilon is some vanishingly small number. Prove that expression and accolades will follow.

Bibliography

[1] M.R. Anderberg. *Cluster Analysis for Applications*. Academic Press, first edition, 1973.

[2] J.C. Aude, Y. Diaz-Lazcoz, J.J. Codani, and J.L. Risler. Applications of the pyramidal clustering method to biological objects. *Comput. Chem.*, 23(3-4):303–315, June 1999.

[3] Open Babel. Open Babel: The Open Source Chemistry Toolbox http://openbabel.org/wiki/Main_Page, January 2010.

[4] RCSB Protein Data Bank. RCSB Protein Data Bank http://www.rcsb.org/pdb/home/home.do, January 2010.

[5] J.M. Barnard and G.M. Downs. Chemical fragment generation and clustering software. *J. Chem. Inf. Model.*, 37(1):141–142, 1997.

[6] S. Bates. Progress towards personalized medicine. *Drug Discovery Today*, November 2009.

[7] P. Bertrand. *Structural Properties of Pyramidal Clustering*, volume 19. American Mathematical Society, 2008.

[8] P. Bertrand. Systems of sets such that each set properly intersects at most one other set-Application to cluster analysis. *Discrete Applied Mathematics*, 156:1220–1236, 2008.

[9] J.C. Bezdek. *Pattern Recognition with Fuzzy Objective Function Algorithms*. Plenum Press, 1981.

[10] BKChem. BKChem http://bkchem.zirael.org, January 2010.

[11] BLAST. BLAST: Basic Local Alignment Search Tool http://blast.ncbi.nlm.nih.gov/Blast.cgi, January 2010.

[12] C. Bohm, B. Braunmuller, F. Krebs, and H. Kriegel. Epsilon grid order: an algorithm for the similarity join on massive high-dimensional data. *SIGMOD Rec*, 30(2):379–388, 2001.

[13] B. Bollobas. *Random Graphs*. Cambridge University Press, second edition, 2001.

[14] C. Bologa, T.K. Allu, M. Olah, M.A. Kappler, and T.I. Oprea. Descriptor collision and confusion: Toward the design of descriptors to mask chemical structures. *J. Comp. Aid. Mol. Design*, 19:625–635, 2005.

[15] A. Bondy and U.S.R. Murty. *Graph Theory*. Springer, first edition, 2008.

[16] I. Borg and P. Groenen. *Modern Multidimensional Scaling*. Springer, second edition, 2009.

[17] J. Bradshaw and R. Sayle. Some thoughts on significant similarity and sufficient diversity http://www.daylight.com/meetings/emug97/ Bradshaw/Significant_Similarity/Significant_Similarity. html, March 1997.

[18] M.L. Brewer. Development of a spectral clustering model for the analysis of molecular data sets. *J. Chem. Inf. Model.*, 47(5):1727–33, 2007.

[19] R. Bro, E. Acar, and T. Kolda. Resolving the sign ambiguity in the singular value decomposition. *J. Chemometrics*, 2008.

[20] R.D. Brown and Y.C. Martin. Use of structure-activity data to compare structure-based clustering methods and descriptors for use in compound selection. *J. Chem. Inf. Comput. Sci.*, 36(3):572–584, 1996.

[21] R.L. Burden and J.D. Faires. *Numerical Analysis*. Brooks Cole, eighth edition, 2004.

[22] D. Butina. Unsupervised data base clustering based on daylight's fingerprint and tanimoto similarity: a fast and automated way to cluster small and large data sets. *J. Chem. Inf. Comput. Sci.*, 39(4):747–750, 1999.

[23] CDK. The Chemistry Development Kit http://sourceforge.net/ projects/cdk, January 2010.

[24] M. Chavent, Y. Lechevallier, and O. Briant. DIVCLUS-T: A monothetic divisive hierarchical clustering method. *Computational Statistics & Data Analysis*, pages 687–701, 2007.

[25] Digital Chemistry. Digital Chemistry - Products http://www. digitalchemistry.co.uk/prod_toolkit.html, January 2010.

[26] X. Chen, A. Rusinko III, and S.S. Young. Recursive partitioning analysis of a large structure-activity data set using three-dimensional descriptors. *J. Chem. Inf. Comput. Sci.*, pages 1054–1062, 1998.

[27] Y. Cheng and G.M. Church. Biclustering of expression data. In *Proceedings of the 8th International Conference on Intelligent Systems for Molecular Biology*, pages 93–103, 2000.

[28] W.S. Cleveland. *Visualizing Data*. Hobart Press, 1993.

[29] R. L. Cook. Stochastic sampling in computer graphics. *ACM Transactions on Graphics*, 5(1):51–72, 1986.

[30] T.H. Cormen, C.E. Leiserson, R.L. Rivest, and C. Stein. *Introduction to Algorithms*. MIT Press, second edition, 2001.

[31] T.F. Cox and M.A.A. Cox. *Multidimensional Scaling, Second Edition*. Chapman & Hall, 2000.

[32] N.E. Day. Estimating the components of a mixture of two normal distributions. *Biometrika*, 56:7–24, 1969.

[33] H. de Fraysseix, J. Pach, and R. Pollack. How to draw a planar graph on a grid. *Combinatorica*, 10:41–51, 1990.

[34] C. Ding, X. He, and H.D. Simon. On the equivalence of nonnegative matrix factorization and spectral clustering. In *Proceedings of the SIAM International Conference on Data Mining*. IEEE, 2005.

[35] S.L. Dixon and R.T. Koehler. The hidden component of size in two-dimensional fragment descriptors: side effects on sampling in bioactive libraries. *J. Med. Chem.*, 42(15):2887–2900, 1999.

[36] R.O. Duda, P.E. Hart, and D.G. Stork. *Pattern Classification*. Wiley Interscience, second edition, 2001.

[37] M.B. Eisen, P.T. Spellan, P.O. Brown, and D. Botstein. Cluster analysis and display of genome-wide expression patterns. In *Proceedings of the National Academy of Science USA*, volume 95, pages 14863–14868. National Academy of Science of the United States of America, 1998.

[38] EMBL-EBI. The EMBL Nucleotide Sequence Database http://www.ebi.ac.uk/embl/, January 2010.

[39] eMolecules. eMolecules Chemical Structure Drawing Search http://www.emolecules.com, January 2010.

[40] P. Erdös and A. Rényi. On random graphs. *Publicationes Mathematicae Debrecen*, 6:290–297, 1959.

[41] B.S. Everitt. *Cluster Analysis*. Wiley, first edition, 1974.

[42] B.S. Everitt, S. Landau, and M. Leese. *Cluster Analysis*. Wiley, fourth edition, 2009.

[43] V. Faber. Clustering and the continuous k-means algorithm. *Los Alamos Sci.*, 22:138–144, 1994.

[44] I. Fáry. On straight line representation of planar graphs. *Acta. Sci. Math. Szeged*, 11:229–233, 1948.

[45] G.S. Fishman. *Monte Carlo Concepts, Algorithms, and Applications.* Springer, first edition, 1996.

[46] M.A. Fligner, J.S. Verducci, and P.E. Blower. A modification of the Jaccard-Tanimoto similarity index for diverse selection of chemical compounds using binary strings. *Technometrics*, 44(2):110–119, 2002.

[47] National Center for Biotechnology Information. National Center for Biotechnology Information `http://www.ncbi.nlm.nih.go`, January 2010.

[48] National Center for Biotechnology Information. PubChem `http:// pubchem.ncbi.nlm.nih.gov`, January 2010.

[49] E.W. Forgy. Cluster analysis of multivariate data: efficiency vs interpretability of classifications. *Biometrics*, 21:768–769, 1965.

[50] D.J. Galas and S.J. McCormack, editors. *Genomic Technologies Present and Future.* Caister Academic Press, 2002.

[51] M.N. Gamito and S.C. Maddock. Accurate multidimensional Poisson-disk sampling. *ACM Transactions on Graphics*, 29(8), 2009.

[52] J.E. Gentle. *Random Number Generation and Monte Carlo Methods.* Springer, second edition, 2004.

[53] A.C.. Gibbs and D.K. Agrafiotis. Chemical diversity: definition and quantification. In P. Bartlett and M. Entzeroth, editors, *Exploiting Chemical Diversity for Drug Discovery*, pages 139–160. The Royal Soceity of Chemistry, 2006.

[54] J.C. Gower and P. Legendre. Metric and Euclidean properties of dissimilarity coefficients. *J. Class.*, 3(3):5–48, 1986.

[55] R.L. Graham, D.E. Knuth, and O. Patashnik. *Concrete Mathematics.* Addison-Wesley Professional, second edition, 1994.

[56] J.A. Grant, M.A. Gallardo, and B.T. Pickup. A fast method of molecular shape comparison: a simple application of a gaussian description of molecular shape. *J. Comb. Chem.*, 17(14):1653–1666, 1996.

[57] J.A. Grant, J.A. Haigh, B.T. Pickup, A. Nicholls, and R.A. Sayle. Lingos, finite state machines and fast similarity searching. *J. Chem. Inf. Model.*, 46(5):1912–1918, 2006.

[58] R. Guha, D. Dutta, D. Wild, and T. Chen. Counting cluster using R-NN curves. *J. Chem. Inf. Model.*, 47(4):1308–1318, 2007.

[59] M. Halkidi, Y. Batistakis, and M. Vazirgiannis. On clustering validation techniques. *J. Intell. Inf. Sys.*, 17(2/3):107–145, 2001.

[60] J. Hammersley. Monte Carlo methods for solving multivariable problems. In *Proceedings of the New York Academy of Science*, volume 86, pages 844–874, 1960.

[61] M. Haranczyk and J. Holliday. Comparison of similarity coefficients for clustering and compound selection. *J. Chem. Inf. Comput. Sci.*, 48(3):819–828, 2008.

[62] J.H. Hardin and F.R. Smietana. Automating combinatorial chemistry: a primer on benchtop robotic systems. *Mol. Div.*, 1(4):270–274, 1996.

[63] J. Hartigan. *Clustering Algorithms*. Wiley, 1975.

[64] J.A. Hartigan. Direct clustering of a data matrix. *J. Am. Stat. Assoc.*, 67(337):123–129, 1972.

[65] J.A. Hartigan. Introduction. In P. Arabie, L.J. Hubert, and G. De Soete, editors, *Clustering and Classification*. World Scientific, 1996.

[66] J.A. Hartigan and M.A. Wong. A K-means clustering algorithm. *Appl. Stat.*, 28:100–108, 1979.

[67] E. Hartuv and R. Shamir. A clustering algorithm based on graph connectivity. *Inf. Proc. Let.*, 76(4-6):175–181, 2000.

[68] T. Hastie, R. Tibshirani, and J. Friedman. *The Elements of Statistical Learning*. Springer, 2001.

[69] S.R. Heller and A.D. McNaught. The IUPAC International Chemical Identifier (InChI). *Chem. International*, 31(1):7, 2009.

[70] J. Herrero, A. Valencia, and J. Dopazo. A hierarchical unsupervised growing neural network for clustering gene expression patterns. *Bioinformatics*, 17:126–136, 2001.

[71] J.D. Holliday, C-Y Hu, and P. Willett. Grouping of coefficients for the calculation of inter-molecular similarity and dissimilarity using 2D fragment bit-strings. *Comb. Chem. & High Throughput Screening*, 11(2):155–166, 2002.

[72] J.D. Holliday, N. Salim, M. Whittle, and P. Willett. Analysis and display of the size dependence of chemical similarity coefficients. *J. Chem. Inf. Comput. Sci.*, 43(3):819–828, 2003.

[73] Daylight Chemical Information Systems Inc. Daylight SMARTS Tutorial http://www.daylight.com/dayhtml_tutorials/languages/smarts/index.html, December 2008.

[74] M. Irwin and B. Shoichet. ZINC—a free database of commercially available compounds for virtual screening. *J. Chem. Inf. Model.*, 45(1):177–82, 2005.

[75] P. Jaccard. Distribution de la flore alpine dans le bassin des Dranses et dans quelques rgions voisines. *Bulletin del la Socit Vaudoise des Sciences Naturelles*, 0(37):241–272, 1901.

[76] A.K. Jain and R.C. Dubes. *Algorithms for Clustering Data.* Prentice Hall, 1988.

[77] D. Jiang, C. Tang, and A. Zhang. Cluster Analysis for Gene Expression Data: A Survey (2004) http://www.cse.buffalo.edu/DBGROUP/ bioinformatics/papers/survey.pdf, January 2004.

[78] Jmol. Jmol:an open-source Java viewer for chemical structures in 3D http://www.jmol.org, January 2010.

[79] M. A. Johnson and G.M. Maggiora, editors. *Concepts and Applications of Molecular Similarity.* Wiley, 1990.

[80] I.T. Jolliffe. *Principal Component Analysis.* Springer-Verlag, second edition, 2002.

[81] P.D. Karp. BioCyc Home http://www.biocyc.org, January 2010.

[82] L. Kaufman and P.J. Rousseeuw. *Finding Groups in Data.* Wiley, 1990.

[83] L.A. Kelly, S.P. Gardner, and M.J. Sutcliffe. An automated approach for clustering an ensemble of NMR-derived protein structures into conformationally related subfamilies. *Prot. Eng.*, 9:1063–1065, 1996.

[84] L. Kocis and W. Whiten. Computational investigations of low-discrepancy sequences. *ACM Trans. on Math. Soft.*, 23(2):266–294, 1997.

[85] A. Kong and et al. Parental origin of sequence variants associated with complex diseases. *Nature*, pages 868–874, December 2009.

[86] B. Kosata. Use of InChI and InChIKey in the XML gold book. *Chem. Internat.*, 31(1), 2009.

[87] H.-P. Kriegel, P. Kröger, and A. Zimek. Clustering high dimensional data: a survey on subspace clustering, pattern-based clustering, and correlation clustering. *ACM Transactions on Knowledge Discovery from Data*, 3(1), 2009.

[88] M. Kull and J. Vilo. Fast approximate hierarchical clustering using similarity heuristics. *BioData Mining*, 1(9), 2008.

[89] M.S. Lajiness and V. Shanmugasundaram. Strategies for the identification and generation of informative compound sets. In J. Bajorath, editor, *Chemoinformatics: Concepts, Methods, and Tools for Drug Discovery*. Springer, 2004.

[90] G. N. Lance and W.T. Williams. A general theory of classificatory sorting strategies: II. Clustering systems. *Comput. J.*, 10:271–277, 1967.

[91] A.R. Leach and V.J. Gillet. *An Introduction to Chemoinformatics*. Kluwer Academic Publishers, first edition, 2003.

[92] P. Legendre and L. Legendre. *Numerical Ecology*. Elsevier, 1998.

[93] Z. Lepp, C. Huang, and T. Okada. Finding key members in compound libraries by analyzing networks of molecules assembled by structural similarity. *J. Chem. Inf. Model.*, 49(11):2429–2443, 2009.

[94] V.I. Levenshtein. Binary codes capable of correcting deletions, insertions, and reversals. *Soviet Physics Doklady*, 10:707–710, 1966.

[95] Y. Li and S.M. Chung. Parallel bisecting k-means with prediction clustering algorithm. *J. Supercomp.*, 39(1):19–37, 2007.

[96] C.A. Lipinksi, F. Lombardo, B.W., Dominy, and P.J. Feeney. Experimental and computational approaches to estimate solubility and permeability in drug discovery and development settings. *Adv. Drug Del. Rev*, 23(23):3–25, 1997.

[97] S.P. Lloyd. Least squares quantization in PCM. *Bell Laboratories, Technical Report*, 1957.

[98] J. MacCuish, C. Nicolaou, and N.E. MacCuish. Ties in proximity and clustering compounds. *J. Chem. Info. Comput. Sci.*, 41(1):134–146, 2001.

[99] J.D. MacCuish, N.E. MacCuish, M. Hawrylycz, and M. Chapman. Quasi-Monte Carlo integration for the generation of molecular shape fingerprints. Manuscript.

[100] J. MacQueen. Some methods for classification and analysis of multivariate observations. In *Proceedings of the Fifth Berkeley Symposium on Mathematical Statistics and Probability*, pages 281–297. University of California Press, 1967.

[101] N. Mantel. The detection of disease clustering and a generalized regression approach. *Cancer Res.*, 27:209–220, 1967.

[102] E.J. Martin, J.M. Blaney, M.A. Siani, D.C. Spellmeyer, A.K. Wong, and W.H. Moos. Measuring diversity: experimental design of combinatorial libraries for drug discovery. *J. Med. Chem.*, 38(9):1431–1436, 1995.

[103] M. Meila and J. Shi. A random walks view of spectral segmentation. In *8th International Workshop on Artificial Intelligence and Statistics*. Society for Artificial Intelligence and Statistics, 2001.

[104] Inc. Mesa Analytics & Computing. SAESAR http://www.mesaac.com, January 2010.

[105] G.W. Milligan. Ultrametric hierarchical clustering algorithms. *Psychometrika*, 44:343–346, 1979.

[106] W.J. Morokoff and R.E. Caflisch. Quasi-Monte Carlo integration. *J. Comput. Phys.*, 122(2):218–230, 1995.

[107] C.A. Nicolaou, J.D. MacCuish, and S.Y. Tamura. A new multi-domain clustering algorithm for lead discovery that exploits ties in proximities. In *13th European Symposium on Quantitative Structure-Activity Relationships*, pages 486–495. Prous Science, 2000.

[108] M.M. Olah, C.C. Bologna, and T.I. Oprea. Strategies for compound selection. *Curr. Drug Disc. Tech.*, 1:211–220, 2004.

[109] C.M. Papadimitriou. *Computational Complexity*. Addison-Wesley, 1992.

[110] A Prinzie and D. Van Den Poel. Incorporating sequential information into traditional classification models by using an element/position-sensitive SAM. *Decision Support Systems*, 42(2):508–526, 2006.

[111] G.J.E Rawlins. *Compared to What*. Computer Science Press, 1992.

[112] ReNaBI. ReNaBI French bioinformatics platforms network http://www.renabi.fr, January 2010.

[113] PIR Protein Information Resource. PIR-PSD Database http://pir.georgetown.edu/pirwww/dbinfo/pir_psd.shtml, January 2010.

[114] G.M. Rishton. Molecular diversity in the context of leadlikeness: compound properties that enable effective biochemical screening. *Curr. Opin. in Chem. Bio.*, 40:340–351, 2008.

[115] G.-C. Rota. The number of partitions of a set. *Amer. Math. Monthly*, 71:498–504, 1964.

[116] N. Salim, J.D. Holliday, and P. Willett. Combination of fingerprint-based similarity coefficients using data fusion. *J. Chem. Inf. Comput. Sci.*, 43(2):435–442, 2003.

[117] W. Schnyder. Embedding planar graphs on the grid. In *Proceedings of the First Annual ACM-SIAM Symposium on Discrete Algorithms*, pages 138–148, 1990.

[118] P.A. Schulte, R.L. Ehrenberg, and M. Singal. Investigation of occupational cancer clusters. *Am. J. Public Health*, pages 52–56, 1987.

[119] R. Shamir and R. Sharan. Click: A clustering algorithm for gene expression analysis. In *Proceedings of the 8th International Conference on Intelligent Systems for Molecular Biology*, 2000.

[120] N.E. Shemetulskis, J.B. Dunbar, B.W. Dunbar, D.W. Moreland, and C. Humblet. Enhancing the diversity of a corporate database using chemical database clustering and analysis. *J. Comp. Aided Mol. Des.*, pages 407–416, 1995.

[121] N.E. Shemetulskis, D. Weininger, C.J. Blankley, J.J. Yang, and C. Humblet. Stigmata: an algorithm to determine structural commonalities in diverse datasets. *J. Chem. Inf. Comput. Sci.*, 36(4):862–871, 1996.

[122] J. Shi and J. Malik. Normalized cuts and image segmentation. *IEEE Trans. on Patt. Anal. and Mach. Intel.*, 22(8):888–905, August 2000.

[123] P. Shirley. Discrepancy as a quality measure for sample distributions. In *Proceedings of Eurographics*, pages 183–194. Association for Computing Machinery, 1991.

[124] S.S. Skiena. *Algorithm Design Manual*. Springer, second edition, 2008.

[125] P.H.A. Sneath. The application of computers to taxonomy. *J. Gen. Microbiol.*, pages 201–226, 1957.

[126] P.H.A. Sneath and R.R. Sokal. *Numerical Taxonomy*. Freeman, first edition, 1973.

[127] R.R . Sokal and P.H.A. Sneath. *Principles of Numerical Taxonomy*. Freeman, first edition, 1963.

[128] M. Steinbach, G. Karypis, and V. Kumar. A comparison of document clustering techniques. In *Proceedings of World Text Mining Conference*. KDD-2000, 2000.

[129] Symyx. Symyx Products Cheminformatics http://www.symyx.com/products/software/cheminformatics/index.jsp, January 2010.

[130] T.T. Tanimoto. Internal Report. *IBM Internal Report*, November 17, 1957.

[131] R. E. Tarjan. A hierarchical clustering algorithm using strongly connected components. *Information Processing Letters*, 14:26–29, 1982.

[132] R.J. Taylor. Simulation analysis of experimental design strategies for screening random compounds as potential new drugs and agrochemicals. *J. Chem. Inf. Comput. Sci.*, 35(1):59–67, 1995.

[133] R Development Core Team. R: A Language and Environment for Statistical Computing http://www.R-project.org, 2008.

[134] RSC Advancing the Chemical Sciences. ChemSpider Building community for chemists http://www.chemspider.com, January 2010.

[135] R. Tibshirani, G. Walther, and T. Hastie. Estimating the number of clusters in a dataset via the gap statistic. *J. Royal. Statist. Soc. B.*, 32(2):411–423, 2001.

[136] E.R Tufte. *The Visual Display of Quantitative Information*. Graphics Press, second edition, 2001.

[137] J.W. Tukey. *Exploratory Data Analysis*. Addison Wesley, 1977.

[138] A. Tversky. Features of similarity. *Psych. Rev.*, 84(4):327–352, 1957.

[139] M.J. Vainio and M.S. Johnson. Generating conformer ensembles using a multiobjective genetic algorithm. *J. Chem. Inf. Model.*, 47:2462–2474, 2007.

[140] M.J. Van der Laan and K.S. Pollard. A new algorithm for hybrid hierarchical clustering with visualization and the bootstrap. *J. Stat. Plan. Inf.*, 117:275–303, 2003.

[141] J.H. van Lint and R.M. Wilson. *A Course in Combinatorics*. Cambridge University Press, 1992.

[142] M.E. Wall, A. Rechtsteiner, and L.M. Rocha. Singular value decomposition and principal component analysis. In D. Berrar, W. Dubitzky, and M. Granzow, editors, *A Practical Approach to Microarray Data Analysis*, pages 91–109. Springer, 2003.

[143] Y. Wang and J. Bajorath. Development of a compound class-directed similarity coefficient that accounts for molecular complexity effects in fingerprint searching. *J. Chem. Inf. Model.*, 49(6):1369–1376, 2009.

[144] J.H. Ward. Hierarchical grouping to optimize and objective function. *J. Amer. Stat. Assoc.*, 58:236–244, 1963.

[145] D. Weininger. SMILES, a chemical language and information system. 1. Introduction to methodology and encoding rules. *J. Chem. Inf. Comput. Sci.*, 28(1):31–36, 1988.

[146] D. Weininger, A. Weininger, and J.L. Weininger. SMILES. 2. Algorithm for generation of unique SMILES notation. *J. Chem. Inf. Comput. Sci.*, 29(2):97–101, 1989.

[147] M. Whittle, P. Willett, W. Klaffke, and P. van Noort. Evaluation of similarity measures for searching the dictionary of natural products database. *J. Chem. Inf. Comput. Sci.*, 43(2):449–457, 2003.

[148] D.J. Wild and C.J. Blankley. Comparison of 2D fingerprint types and hierarchy level selection methods for structural grouping using wards clustering. *J. Chem. Inf. Comput. Sci.*, 40:155–162, 2000.

[149] P. Willett. *Similarity and Clustering in Chemical Information Systems.* Research Studies Press, first edition, 1987.

[150] P. Willett, J.M. Barnard, and G.M. Downs. Chemical similarity searching. *J. Chem. Inf. Comput. Sci.*, 38:983–996, 1998.

[151] R.J Wilson and J.J. Watkins. *Graphs An Introductory Approach.* John Wiley & Sons, first edition, 1990.

[152] W.J. Wiswesser. *A Line-Formula Chemical Notation.* Crowell Co., New York edition, 1954.

[153] L. Xue, J.W. Godden, and J. Bajorath. Evaluation of descriptors and mini-fingerprints for the identification of molecules with similar activity. *J. Chem. Inf. Comput. Sci.*, 40(5):1227–1234, 2000.

[154] K.Y. Yeung and W. L. Ruzzo. Principal component analysis for clustering gene expression data. *Bioinformatics*, 17(9):763–774, 2001.

[155] L.A. Zadeh. Fuzzy sets. *Inf. Control*, 8(3):338–353, 1965.

[156] R. Zass and A. Shashua. A unifying approach to hard and probabilistic clustering. In *Proceedings of the International Conference on Computer Vision (ICCV) Beijing*. IEEE, 2005.

[157] C.T. Zhang, K.C. Chou, and G.M. Maggiora. Predicting protein structural classes from amino acid composition: application of fuzzy clustering. *Prot. Eng.*, 8(5):425–435, 1995.

[158] M. Zvelebil and J. Baum. *Understanding Bioinformatics.* Garland Science, first edition, 2008.

Index

Milton Keynes UK
Ingram Content Group UK Ltd.
UKHW031148141024
449569UK00024B/988

9 781138 374232